THE ZOOLOGY *of the* VOYAGE OF H.M.S. BEAGLE.

UNDER THE COMMAND
OF CAPTAIN FITZROY.
DURING THE YEARS
1832 TO 1836.

FOSSIL MAMMALIA
PART I

By

RICHARD OWEN, ESQ. F.R.S

Edited and Superintended by

CHARLES DARWIN,
ESQ. M.A. F.R.S. SEC. G.S.

First published in 1840

This edition published by Read Books Ltd.
Copyright © 2019 Read Books Ltd.
This book is copyright and may not be
reproduced or copied in any way without
the express permission of the publisher in writing

British Library Cataloguing-in-Publication Data
A catalogue record for this book is available
from the British Library

CONTENTS

Charles Darwin 9

Preface. 13

Geological Introduction. 19

Fossil Mammalia. 31

A Description of the Cranium of Toxodon Platensis; 34

Description of Parts of the Skeleton of Macrauchenia
Patachonica; 61

Description of a Fragment of a Cranium of an Extinct
Mammal, Indicative of a New Genus of Edentata, and
for Which is Proposedthe Name of Glossotherium. 93

Description of a Mutilated Lower Jaw and Teeth, on
Which is Founded a Subgenus of Megatherioid
Edentata, Under the Name of Mylodon. 102

Description of a Considerable Part of the Skeleton of a
Large Edentate Mammal, allied To the Megatherium
and Orycteropus, and for Which is Proposed the Name
of Scelidotherium Leptocephalum. 116

Description of a Mutilated Lower Jaw of the
Megalonyx Jeffersonii. 153

Description of a Fragment of the Skull and of the
Teeth of themegatherium Cuvieri. 154

Description of Fragments of Bones, and of Osseous
Tesselated Dermal Covering of Large Edentata. 164

Notice of Fragments of Molar Teeth of a Mastodon. 167

Notice of the Remains of a Species of Equus, ... 167

Description of Remains of Rodentia, Including the
Jaws and Teeth of an Extinct Species of Ctenomys. 169

ILLUSTRATIONS

Base of the Skull of Toxodon Platensis. 173

Side View of the Skull of Toxodon 174

Top View of the Skull of the Toxodon. 175

6th Grinder—Upper Jaw Toxodon Platensis. 176

Fragments of the lower Jaw and Teeth of a Toxodon. 177

Cervical Vertebræ of Macrauchenia. 178

Cervical Vertebræ of 1.2. Macrauchenia. 3.4. Auchenia. . . 179

Lumbar Vertebræ; Macrauchenia. 180

Macrauchenia. 181

Proximal Extremity of anchylosed Ulna and Radius
Macrauchenia. 182

Bones of the right fore-foot, Macrauchenia. 183

Right Femur. Macrauchenia. 184

Macrauchenia. Right Tibia and Fibula. 185

Right Astragalus. Macrauchenia. 186

Macrauchenia. Fig: 1 Metatarsal. 2-5. Metacarpals. 187

Fragment of the Cranium of the Glossotherium. 188

1. Megalonyx Jeffersoni. 2. Meg laqueatus. 3,4 Mylodon
Harlani. 5. Myl Darwinii. 189

Mylodon. 190

Mylodon.. 191

Scelidotherium. *192*

Scelidotherium. *193*

Scelidotherium. *194*

Cranial Cavity and Dentition of Scelidotherium. *195*

Cervical and Anterior dorsal Vertebræ. *196*

Scelidotherium. *197*

Left Astragalus. *198*

Scelidotherium. *199*

Left Astragalus.. .... *200*

Lower Jaw of Megalonyx. *201*

Megatherium.. .... *202*

Section of the superior maxillary teeth, Megatherium. . . . *203*

1. Megatherium. 2-5 Hoplophorus. 6-12. Ctenomys. 13-14. Equus. *204*

Charles Darwin

Charles Robert Darwin was born on 12 February 1809, in Shrewsbury, Shropshire, UK. He is best known for his pioneering work in evolutionary theory; establishing that all species of life have descended from common ancestors through a process of 'natural selection'. This theory of evolution was published in the 1859 book, *On the Origin of Species,* a text which has become a seminal work of modern science.

Darwin was the fifth of six children of the wealthy society doctor and financier Robert Darwin, and his wife Susannah Darwin (*née* Wedgwood). He was the grandson of two prominent abolitionists: Erasmus Darwin on his father's side, and Josiah Wedgwood on his mother's side. Darwin spent his early education in the local Shrewsbury School as a boarder, before moving to Edinburgh in 1825 to study medicine at the University. Darwin's early interest in nature led him to neglect his medical studies however. He found the medical lectures dull, and the surgery distressing. Instead, he helped investigate marine invertebrates in the Firth of Forth with Robert Edmond Grant. This neglect of medical studies annoyed his father, who sent Darwin to *Christ's College, Cambridge*, in order to undertake a Bachelor of Arts degree - as the first step towards becoming an Anglican parson. As Darwin was unqualified for the *Tripos*, he joined the *ordinary* degree course in January 1828. He successfully graduated in 1831, but continued his investigations into the natural world, particularly partaking in the popular craze for beetle collecting. On his graduation, Darwin was invited to join the voyage of the HMS Beagle, with Captain Robert FitzRoy - a journey which lasted almost five years and traversed the globe. The journal of this voyage on the HMS Beagle (published in 1839) established Darwin as a popular author; he detailed his time spent investigating geology and making natural history collections whilst on land. He kept careful notes of his observations and theoretical speculations, and at intervals during the voyage his specimens were sent to *Cambridge University.* When the *Beagle* reached Falmouth, Cornwall, on 2 October 1836, Darwin was already a celebrity in scientific circles. Puzzled by the geographical distribution of wildlife and fossils he collected on the voyage, Darwin began detailed investigations in 1838 – leading to the conception of his theory of natural selection. Although he discussed his ideas with several naturalists, Darwin needed time for extensive research and his geological work had priority. He was in the process of writing up his theory in 1858 when Alfred Russell Wallace sent him an essay which described the same idea, prompting the immediate joint publication of both of their papers.

Despite repeated bouts of illness during the last twenty-two years of his life, Darwin's work continued. Having published *On the Origin of Species* as an abstract of his theory in 1859, he pressed on with experiments, research, and the writing of what he saw as his magnum opus. *The Variation of Animals and Plants under Domestication* of 1868 was the first part of Darwin's planned 'big book', and included his unsuccessful hypothesis of pangenesis; an attempt to explain heredity. It was a moderate commercial success and was translated into many languages. This was followed by a second part, on natural selection, but it remained unpublished in his lifetime.

Darwin also examined human evolution in specific, and wrote on sexual selection in *The Descent of Man, and Selection in Relation to Sex* (1871). This text was shortly followed by *The Expression of the Emotions in Man and Animals,* and a series of books on botany, including *Insectivorous Plants, The Effects of Cross and Self Fertilisation in the Vegetable Kingdom,* and *The Power of Movement in Plants.* In his last book he returned to *The Formation of Vegetable Mould through the Action of Worms* (1881). By this time, Darwin's health was failing however, and in 1882, he was diagnosed with 'angina pectoris'; a disease of the heart. Darwin died shortly after this diagnosis, at Down House, Kent, on 19 April 1882, and was honoured with a major ceremonial funeral. He is buried at Westminster Abbey, close to John Herschel and Isaac Newton. As a result of his scientific work, Darwin has been described as one of the most influential thinkers in history.

PREFACE.

HIS MAJESTY'S ship, Beagle, under the command of Captain FitzRoy, was commissioned in July, 1831, for the purpose of surveying the southern parts of America, and afterwards of circumnavigating the world. In consequence of Captain FitzRoy having expressed a desire that some scientific person should be on board, and having offered to give up part of his own accommodations, I volunteered my services; and through the kindness of the hydrographer, Captain Beaufort, my appointment received the sanction of the Admiralty. I must here, as on all other occasions, take the opportunity of publicly acknowledging with gratitude, the obligation under which I lie to Captain FitzRoy, and to all the Officers on board the Beagle, for their constant assistance in my scientific pursuits, and for their uniform kindness to me throughout the voyage. On my return (October, 1836) to England, I found myself in possession of a large collection of specimens in various branches of natural history; but from the great expense necessary to secure their publication, I was without the means of rendering them generally serviceable.

The Presidents of the Linnean, Zoological, and Geological Societies, having given me their opinion respecting the utility to be derived from publishing these materials, I addressed a letter to the Right Honourable the Chancellor of the Exchequer (T. Spring Rice, Esq.) informing him of the circumstances under which I hoped that I might venture to solicit the aid of Government. In reply, I received a communication (as below) announcing to me that the Lords of the Treasury, from their readiness to promote Science, were willing, under certain conditions, to give me the most liberal assistance.

"*Treasury Chambers, August* 31, 1837.

"Sir,

"It having been represented to the Lords Commissioners of Her Majesty's Treasury, from various quarters, that great advantage would be derived to the Science of Natural History, if arrangements could be made for enabling you to publish, in a convenient form, and at a cheap rate, the result of your labours in that branch of science, my Lords will feel themselves justified in giving their sanction to the application of a sum, not exceeding in the whole one thousand pounds, in aid of such a publication; upon the clear and distinct understanding that the Work should be published, and the plates engraved, in such a manner as to be most advantageous to the Public at large, upon a plan of arrangement to be previously submitted to, and sanctioned by the Board, after consultation with those persons, who, from their attainments in this branch of science, are the most capable of advising their Lordships thereupon; and that the payments on account of the said sum of one thousand pounds are to be made to you from time to time, on a certificate that such progress has been made in the engravings, in accordance with the plan previously approved of, as to justify the issue then applied for. My Lords have therefore directed me to communicate to you the views they entertain upon this subject; and to apprize you that they will be prepared to act in conformity with their arrangement, upon learning from you that you are ready to proceed with the Work upon the principles above laid down, and upon receiving from you a statement of the manner in which you think the Work should be published, and the plates engraved, so as most effectually to accomplish the object my Lords have in view, in sanctioning the payment from the Public Funds, in aid of the expenses of the Work in question.

"I remain,
Sir, Your Obedient Servant,

"A. Y. SPEARMAN."

The object of the present Work is to give descriptions and figures of undescribed and imperfectly known animals, both fossil and recent, together with some account, in the one case, of their geological position, and in the other of their habits and ranges. As I do not possess the knowledge requisite for such an undertaking, and as I am, moreover, particularly engaged in preparing an account of the geological observations, made during the voyage, several gentlemen have most kindly undertaken different portions of the Work. Besides the very great advantage insured in thus enlisting the attainments of these Naturalists in the several departments of science, to which they have paid most attention, a great delay is avoided by adopting this method of publication, which must otherwise have been incurred before the materials could have been made known.

An Account of the Voyage, drawn up by Captain FitzRoy, (and to which I have added a volume) being on the point of publication, I shall not in this Work enter on any minute details respecting the countries which were visited, but shall merely give a sketch of the geology in the introduction to the part containing Fossil Mammalia, and a brief geographical notice in that attached to the account of existing animals. At the conclusion of this Work, I shall endeavour to place together the leading results in the natural history of the different countries, from which the collections were procured. I may here state that Mr. Owen has undertaken the description of the Fossil Mammalia; Mr. Waterhouse, the Recent Mammalia; Mr. Gould, the Birds; Mr. Bell, the Reptiles; and the Rev. L. Jenyns, the Fish. Whatever assistance I may obtain in the invertebrate classes, will be noticed in their respective places. The specimens have been presented to the various public museums, in which it was thought they would be of most general service: mention will be made in each part where the objects described have been deposited.

FOSSIL MAMMALIA

Described by

RICHARD OWEN,
ESQ. F.R.S. F.G.S. F.L.S.

Professor of Anatomy and Physiology to The Royal College of Surgeons in London; Corresponding Member of The Royal Academy of Sciences of Berlin; of The Royal Academy of Medicine, and Philomathic Society of Paris; of The Academy of Sciences of Philadelphia, Moscow, Erlangen.

WITH

A GEOLOGICAL INTRODUCTION,

BY

CHARLES DARWIN,
ESQ. M.A. F.G.S. &c.&c.

Corresponding Member of The Zoological Society.

GEOLOGICAL INTRODUCTION.

BY MR. DARWIN.

MR. OWEN having undertaken the description of the fossil remains of the Mammalia, which were collected during the voyage of the Beagle, and which are now deposited in the Museum of the College of Surgeons in London, it remains for me briefly to state the circumstances under which they were discovered. As it would require a lengthened discussion to enter fully on the geological history of the deposits in which these remains have been preserved, and as this will be the subject of a separate work, I shall here only give sufficient details, for the reader to form some general idea of the epoch, at which these animals lived,—of their relative antiquity one to the other,—and of the circumstances under which their skeletons were embedded. All the remains were found between latitudes 31° and 50° on the eastern side of South America. The localities may conveniently be classed under three divisions, namely—the Provinces bordering the Plata; Bahia Blanca situated near the confines of Northern Patagonia; and Southern Patagonia.

The first division includes an enormous area, abounding with the remains of large animals. To the eastward and southward of the great streams, which unite to form the estuary of the Plata, those almost boundless plains extend, which are known by the name of the Pampas. Their physical constitution does not vary over a wide extent;—the traveller may pass for many hundred miles on a level surface, without meeting with a single pebble, or discovering any change in the nature of the soil. The formation consists of a reddish argillaceous earth, generally containing irregular concretions of a pale brown, indurated marl. This stone, where most compact, is traversed by small linear cavities, and in several respects resembles the less pure fresh-water limestones of Europe. The concretions not unfrequently become so numerous, that they unite and form a continuous stratum, or even the entire mass.

At Bajada de Sta. Fé, in the Province of Entre Rios, beds of sand, limestone, and clay of different qualities, containing sharks' teeth and sea-shells, underlie the Pampas deposit. The shells, although numerous, are few in kind. Mr. George B. Sowerby informs me that they appear to belong to one of the less ancient tertiary epochs; they consist of *Venus nov. spec.* near to *V. cancellata*; *Arca nov. spec.* near to *A. antiquata*; a very large oyster, probably an extinct species; an imperfect specimen of a second species of oyster near to *O. edulis*; and a *Pecten* near to *P. opercularis*. These beds pass upwards into an indurated marl, and this again into the red argillaceous earth of the Pampas, containing the remains of those extinct quadrupeds, which every where characterize that deposit. To the southward of the Plata level plains of an uniform composition, interrupted only at wide intervals by hills of crystalline rock, extend to a distance of about three hundred miles; and to the northward for at least an equal space, and probably much further. As might have been expected from the perfectly level surface, wherever a continuous section is presented on the banks of the great rivers, very slight changes of colour show, that the deposit has been accumulated in strata as horizontal as the land, or as the water-line at the base of the cliffs.

In the province of Banda Oriental (to the N. and N. E. of the Plata), and in part of that of Entre Rios, the land, though very low and level, has a foundation of granitic and other primary rocks. These older formations are partially covered, in most parts, by a reddish earthy mass containing a few small calcareous concretions; while in other parts, they are concealed by more regular strata, of indurated marl passing into limestone, of conglomerates, and ferruginous sandstone. The entire formation probably belongs to the same epoch with that of the Pampas deposit. In the earthy mass, even where it is of little thickness, and where it might readily be mistaken for detritus produced from the underlying granites, remains of large quadrupeds have several times been discovered.

On the shores of the Plata and in the neighbouring districts, proofs of a change of level having taken place between the land and the water within a recent period, may be observed. Both near Monte Video and Colonia del Sacramiento, beds of shells are lying on the beach at the height of several feet above the present tidal action. Near Maldonado I saw estuary shells of recent species embedded in clay, and raised above the level of a neighbouring fresh-water lake.

On the banks of the Parana, a shell identical with, or most closely resembling an estuary species (*Potamomya labiata*, now living in that part of the Plata, where the water is brackish) is accumulated in great masses, which are found some miles inland, and are elevated several yards above the level of the river. Sir Woodbine Parish, also, has in his possession, shells procured from an extensive formation near Ensenada de Barragan (south of Buenos Ayres), which is quarried for lime. Mr. George Sowerby has examined these fossils, and says the following are identical with living kinds; *Voluta colocynthis*, Dillwyn: *V. angulata*, Swainson: *Buccinum globulosum*, Kiener: a variety of *Oliva patula*: a *Cytherœa* closely resembling or identical with *C. flexuosa*, and a fragment of a second species, probably *C. purpurascens*; *Potamomya labiata*; and fragments of oysters. There is, however, a species of *Mactra* in very great numbers, with which Mr. Sowerby is wholly unacquainted. I may observe that I found recent shells of the first five species inhabiting the coast, a short distance to the southward. Some shelly limestone from the same place, which Sir Woodbine Parish had the kindness to show me, resembles that which I saw at Bajada, and in Banda Oriental. These beds, therefore, probably form parts of the Pampas deposit, and are not merely indicative of the period of its elevation. Nevertheless, on the opposite shores of the Plata, near the mouth of the Uruguay, I found lines of sand dunes, where the *Mactra* and *Cytheræa flexuosa* were lying in such quantities on the bare surface, that the inhabitants, by merely sifting the sand, collect them for burning into lime.

After these facts we may feel certain, that at a period not very remote, a great bay occupied the area both of the Pampas and of the lower parts of Banda Oriental. Into this bay the rivers which are now united in the one great stream of the Plata, must formerly have carried down (as happens at the present day) the carcasses of the animals, inhabiting the surrounding countries; and their skeletons would thus become entombed in the estuary mud which was then tranquilly accumulating. Nothing less than a long succession of such accidents can account for the vast number of remains now found buried. As their exposure has invariably been due to the intersection of the plain by the banks of some stream, it is not making an extravagant assertion, to say, that any line whatever drawn across the Pampas would probably cross the skeleton of some extinct animal.

At Bajada, a passage, as I have stated, may be traced upwards from the beds containing marine shells, to the estuary mud with the bones of land animals. In another locality a bed of the same mineralogical nature with the Pampas deposit, underlies clay containing large oysters and other shells, apparently the same with those at Bajada. We may, therefore, conclude that at the period when the Arca, Venus, and Oyster were living, the physical condition of the surrounding country was nearly the same, as at the time when the remains of the mammalia were embedded; and therefore that these shells and the extinct quadrupeds probably either co-existed, or that the interval between their respective existences was, in a geological point of view, extremely short. In this part of South America there is reason to believe that the movements of the land have been so regular, that the period of its elevation may be taken as an element in considering the age of any deposit. The circumstance, therefore, that the beds immediately bordering the Plata, contain very nearly the same species of molluscs, with those now existing in the neighbouring sea, harmonizes perfectly with the more ancient (though really modern) tertiary character of the fossils underlying the Pampas deposit at Bajada, situated at a greater height, and at a considerable distance in the interior. I feel little doubt that the final extinction of the several large quadrupeds of La Plata did not take place, until the time when the sea was peopled with all, or nearly all, its present inhabitants.

Bahia Blanca, situated in latitude 39°, and about 250 miles south of the Plata, constitutes the second district, in which I found the remains of quadrupeds. This large bay is nearly surrounded by very low land, on which successive lines of sand dunes mark in many parts the retreat of the water. At some distance inland a formation of highly indurated marl, passing into limestone, forms an escarpment. Beyond this, rocks of the same character extend over a wide and desolate plain, which rises towards the flanks of the distant mountain of the Sierra de la Ventana, composed of quartz. On the low shores of this bay, only two places occur, where any section of the strata can be seen; and at both of these I found fossil remains.

At Monte Hermoso, a line of cliff of about 120 feet in height, consists in the upper part of a stratum of soft sandstone with quartz pebbles; and in the lower of a red argillaceous earth, containing concretions of pale indurated marl. This lower bed has the same mineralogical character with the Pampas deposit; and possibly may be connected with it. The embedded bones were blackened, and had undergone more chemical change than in any other locality, which I examined. With the exception of a few large scattered bones, the remains seemed to belong chiefly to very small quadrupeds.

In another part of the bay, called Punta Alta, about eighteen miles from Monte Hermoso, a very small extent of cliff, about twenty feet high, is exposed. The lower bed seen at ebb tide, extends over a considerable area; it consists of a mass of quartz shingle, irregularly stratified, and divided by curved layers of indurated clay. The pebbles are cemented together by calcareous matter, which results, perhaps, from the partial decomposition of numerous embedded shells. In this gravel the remains of several gigantic animals were extraordinarily numerous. The cliff, in the part above high-water mark, is chiefly composed of a reddish indurated argillaceous earth; which either passes into, or is replaced by, the same kind of gravel, as that on which the whole rests. The earthy substance is coarser than that at Monte Hermoso, and does not contain calcareous concretions. I found in it a very few fragments of shells, and part of the remains of one quadruped.

From the bones in one of the skeletons, and likewise from those in part of another, being embedded in their proper relative positions, the carcasses of the animals, when they perished, were probably drifted to this spot in an entire state. The gravel, from its stratification and general appearance, exactly resembles that which is every day accumulating in banks, where either tides or currents meet; and the embedded shells are of littoral species. But from the skeleton, in one instance, being in a position nearly undisturbed, and from the abundance of serpulæ and encrusting corallines adhering to some of the bones, the water, at the time of their burial, must have been deeper than at present. This conclusion might also have been inferred from the fact, that in the neighbouring cliff the same bed, with its shells, has been uplifted some yards above high-water mark. On the coast to the southward abundant proofs occur, of a recent elevation of the continent. In the gravel, nearly all the pebbles are of quartz, and have originally proceeded from the lofty range of the Ventana, distant between forty and fifty miles. Besides the pebbles of quartz, there are a few irregular masses of the same indurated marl, of which the escarpment of the neighbouring great plain is composed. Hence the gravel beds must have been deposited, when the plain existed as dry land; and on it probably those great animals once lived, of which we now find only the remains. The indurated marl forming the plain, is the same kind of rock with that occurring over a wide extent of the Pampas; and there is no reason to doubt, they are parts of one great formation. Nevertheless, the gravel bed of Bahia Blanca, although subsequent to the calcareous formation, may be of the same age with those parts of the Pampas, which stand at a low level near the Plata. For on this whole line of coast, I believe, as the land has continued rising, fresh littoral deposits have been formed; and each of these would often owe part of its materials to the degradation of the one last elevated.

With respect to the relative age of the Monte Hermoso and Punta Alta beds, it is not possible to speak decidedly. A certain degree of similarity in the nature of the strata containing quartz pebbles, and those of the reddish indurated earth; and the short distance between the two localities, would indicate that no long interval had intervened. The beds at Monte Hermoso, certainly were deposited more tranquilly, and probably in a deeper sea; so that even skeletons of animals, no larger than rats, have been perfectly preserved there. In some parts of the surrounding country, obscure traces of a succession of step-formed terraces may be observed; and each of these indicates a period of repose during the elevation of the land, at which time the strata previously existing were worn away, and fresh matter deposited. The Monte Hermoso beds were, perhaps, formed during one such interval, anterior to the accumulation of the shingle bank at Punta Alta.

Mr. G. Sowerby, who has been good enough to examine the shells which were found with the remains of the quadrupeds, has given me the following list.

1. *Voluta angulata.*
2. —— *colocynthis.*
3. *Oliva Brasiliensis.*
4. ——   Nearly related to *O. patula*, but specimen imperfect.
5. ——   Nearly related to *O. oryza*; less nearly to small species now living at Bahia Blanca.
6. ——   *Nov. spec.*
7. *Buccinum cochlidium.*
8. ————*globulosum.*
9. —————   One or two minute species, perhaps young specimens, — unknown.
10. *Trochus*   *Nov. spec.* (?) same as one now living in the bay.
11. ———   *Nov. spec.* (?) nearly related to last; differs in not being granular on the surface.
12. *Assiminia* (?)   Minute species, identical with one living in the bay.
13. *Bulinus nucleus.*
14. *Fissurella*   Probably same as a kind (*nov. spec.* ?) living in the bay.
15. *Crepidula muricata.*
16. —————   *Nov. spec.*
17. *Cytherœa*   Closely related to, or identical with *C. purpurascens.*
18. *Modiola*   Same as recent kind (*nov. spec.*) living in the bay.
19. *Nucula*   Near to *N. margaritacea.*
20. *Corbula*   Minute species, unknown.
21. *Cardita*   Ditto   ditto
22. *Pecten*   *Nov. spec.* (?) very imperfect specimen.
23. *Ostrea*   Oysters of the same size now live in the bay.

I may add that a fossil encrusting coralline is the same with one now living in the bay.

Of these shells it is almost certain that twelve species (and the coralline) are absolutely identical with existing species; and that four more are perhaps so; the doubt partly arising from the imperfect condition of the specimens. Of the seven remaining ones, four are minute, and one extremely imperfect. If I had not made a collection (far from perfect) of the shells now inhabiting Bahia Blanca, Mr. Sowerby would not have known as living kinds, five out of the twelve fossils: therefore, it is probable, if more attention had been paid to collecting the small living species, some of the seven unknown ones would also have been found in that state. The twelve first shells, as well as the four doubtful ones, are not only existing species, but nearly all of them inhabit this same bay, on the shores of which they are likewise found fossil. Moreover, at the time, I particularly noticed that the proportional numbers appeared closely similar between the different kinds,—in those now cast up on the beach, and in those embedded with the fossil bones. Under these circumstances, I think, we are justified (although some of the shells are at present unknown to conchologists) in considering the shingle strata at Punta Alta, as belonging to an extremely modern epoch.

From the principle already adduced, namely, the regular and gradual elevation of this part of the continent, I should have judged from the small altitude of the beds at Punta Alta, that the formation had not been very ancient. The conclusion here arrived at, concerning the age of these fossil mammalia, is nearly the same, with that, inferred respecting those entombed in the Pampas; and it will hereafter be shown, that some of the species are common to the two districts. We may suppose, that whilst the ancient rivers of the Plata occasionally carried down the carcasses of animals existing in that country, and deposited them in the mud of the estuary; other animals inhabited the plains round the Sierra de la Ventana, and that lesser streams, acting together with the currents of a large bay, drifted their remains towards a point, where sand and shingle were accumulating into a shoal. The whole area has since been elevated: the estuary mud of the former rivers has been converted into wide and level plains, and the shoals of the ancient Bahia Blanca now form low headlands on the present coast.

The third locality, which I have to specify, is Port St. Julian, in latitude 49° 15' on the coast of Southern Patagonia. The tertiary plains of that country are modelled into a succession of broad and level terraces, which abut one above the other; and where they approach the coast, are generally cut off by a line of precipitous cliff. The whole surface is thickly covered by a bed of gravel, composed of various kinds of porphyries, and probably originating from rocks situated within the Cordillera. The lower part of the formation consists of several varieties of sandstone, and contains many fossil shells, the greater number of which are not found in a living state.

The south side of Port St. Julian is formed by a spit of flat land, of nearly a hundred feet in height; and on its surface existing species of littoral shells are abundantly scattered. The gravel is there covered (a circumstance which I did not observe in scarcely any other locality) by a thin but irregular bed of a sandy or loamy soil, which likewise fills up hollows or channels worn through it. In the largest of these channels the remains of the single fossil quadruped, which was here discovered, were embedded. The skeleton probably was at first perfect; but the sea having washed away part of the cliff, has removed many of the bones,—the remaining ones, however, still occupying their proper relative position to each other. I am inclined to attribute the origin of this earthy matter, to the mud which might have accumulated in channels, and on the surface of the gravel, if this part of the plain had formerly existed as a harbour, such as Port St. Julian is at the present day. The Guanaco, the only large animal now inhabiting the wild plains of Patagonia, often wanders over the extensive flats, which are left dry at the head of the harbour during ebb tide: we may imagine that the fossil animal, whilst in a like manner crossing the ancient bay, fell into one of the muddy creeks, and was there buried.

I have stated that existing species of shells are scattered over the surface of this plain; namely, *Mytilus Magellanicus*; a second and undescribed species, now living on the beach; *M. edulis*; *Patella deaurata*; and on another part of the coast, but having similar geological relations, *Fusus Magellanicus*; *Voluta ancilla*; and a *Balanus*:—all these shells are among the commonest now living on this coast. Although they must have been lying exposed to the atmospheric changes for a very long period, they still partially retain their different colours. From these facts we know, with certainty, that the superficial deposit, containing the remains of the quadruped, has been *elevated* above the sea, within the recent period. From the structure of the step-like plains, which front the coast, it is certain that each step must have been modelled, subsequently to the elevation of the one standing above it; and, as the same recent shells occur on two higher plains, we may, with safety, conclude, that the earthy matter, forming the surface of this lower one, together with its embedded skeleton, was *deposited* long after the existence of the present species, still inhabitants of the sea. According, therefore, to the chronology, taken from the duration of species among the molluscs, the fossil quadruped of Port St. Julian must have been coeval, or nearly so, with those from Bahia Blanca.

Having now briefly described the principal circumstances in the geology of the three districts, to which I at first alluded, I will conclude, by observing, that the fossil mammalia of La Plata, Bahia Blanca, and Port St. Julian, must all have lived during a very modern period in the geological history of the world. It is not the proper place in this work to enter on any speculations, concerning the cause of the extinction of so many gigantic animals. I will only here add, that there is the strongest evidence against admitting the theory of a period of overwhelming violence, by which the inhabitants of the land could have been swept away, and destroyed. On the contrary every thing indicates a former state of tranquillity, during which various deposits were accumulating near the then existing coasts, in the same manner, as we may suppose others are at this day in progress. The only physical change, which we know has taken place, since the existence of these ancient mammalia, has been a small and gradual rising of the continent; but it is difficult to believe, that this alone could have so greatly modified the climate, as to have been the cause of the utter extermination of so many animals. Mr. Owen will mention the exact locality where the remains of each quadruped were discovered; and, at the conclusion, it will be easy to specify by name those, which, from being embedded in the same deposit, are known formerly to have coexisted on the continent of South America.

FOSSIL MAMMALIA.

BY MR. OWEN.

IT may be expected that the description of the osseous remains of extinct Mammalia, which rank amongst the most interesting results of Mr. Darwin's researches in South America, should be preceded by some account of the fossil mammiferous animals which have been previously discovered in that Continent. The results of such a retrospect are, however, necessarily comprised in a very brief statement; for the South American relics of extinct Mammalia, hitherto described, are limited, so far as I know, to three species of Mastodon, and the gigantic Megatherium.

One of the above species of Mastodon *(Mast. Cordillerarum)* was established by Cuvier[*] on remains discovered by Humboldt, in Quito, near the volcanic mountain, called *Imbaburra*, at an elevation of 1200 toises above the level of the sea; and likewise at the Cordilleras of Chiquitos, near Santa Cruz de la Sierra, a locality which is near the centre of South America. A second species *(Mastodon Humboldtii,* Cuv.[†]*)* is indicated by molar teeth, stated to have been discovered by the same philosophic traveller, in Chile, near the city of Concepcion. The third species of Mastodon appears to have once ranged in vast troops over the wide empire of Peru: numerous teeth were brought thence to Paris by Dombey,[‡] and similar teeth, together with a humerus and tibia from Santa Fé de Bogota were placed by Humboldt at the disposal of Cuvier,[§] who considered them to belong to the

---

[*] See Ossemens Fossiles, Ed. iv. tom. ii. p. 368. Pl. 27. fig 1. 12.
[†] *Ibid.* p. 370. Pl. 27. fig. 5.
[‡] *Ibid.* p. 347, 367.

[§] *Ibid.* p. 337. Pl. 26. fig. 7.

Mastodon angustidens, a species of which the fossil remains are by no means uncommon in several localities of Europe. Cuvier is also disposed to refer to the same species the teeth of the Mastodon from Brazil and Lima, mentioned by Dr. W. Hunter in his observations on the *animal incognitum* from the Ohio.* The Megatherium has been scientifically described and illustrated in the works of Bru, Cuvier, and D'Alton, whose accounts are founded on a nearly complete skeleton of this stupendous quadruped which has existed in the Royal Museum at Madrid for more than half a century. The few deficiencies in its osteography have recently been supplied by the descriptions and figures given by Dr. Buckland† and Mr. Clift,‡ taken from remains of the Megatherium, brought by Sir Woodbine Parish from Buenos Ayres, and which were discovered in the bed of the Rio Salado, a tributary of the Rio Plata. Sir Woodbine Parish's collection from the same locality, includes also remains of other species of extinct Edentata, which have not yet been described. M. D'Orbigny, in his travels in South America (vol. i. p. 96.), states that, in the banks of the Parana, he found the fossil remains of a large quadruped, of the size of an Ox, — another quadruped of the size of a Cat, apparently of the carnivorous order;—and a third, a Rodent as large as a Rat.

This meagre condition of the historical part of the subject of South American fossils by no means arises from their actual scarcity. The writings of some of the old Spanish authors, for instance, Torrubia, Garcillasso, and others,§ contain frequent allusions to the bones of giants, who in times of old dwelt in Peru. Legentil, also, in 1728, speaks as an eye-witness of these Peruvian remains; and his guides pointed out to him the traces of the thunder-bolts, by which the Anaks of the New World had been exterminated. Bones and teeth of the Mastodon are,

---

* Philosophical Transactions, vol. lviii. p. 34. (1768).
† Bridgewater Treatise, p. 139.
‡ Geological Transactions, vol. iii. p. 437. pl. 44, 45, 46.
§ Quoted by Cuvier, Ossem. Foss. Ed. iv. tom. ii. p. 351.

according to Humboldt, so abundant in a locality near Santa Fé de Bogota in Columbia, that to this day it bears the name of the "Field of Giants."

But independently of these indications, the abundance and variety of the osseous remains of extinct Mammalia in South America are amply attested by the materials for the following descriptions, collected by one individual, whose

sphere of observation was limited to a comparatively small part of South America; and the future traveller may fairly hope for similar success, if he bring to the search the same zeal and tact which distinguish the gentleman to whom Oryctological Science is indebted for such novel and valuable accessions.

It is remarkable that all the fossils, collected by Mr. Darwin, belong to herbivorous species of mammalia, generally of large size. The greater part are referrible to the order which Cuvier has called Edentata, and belong to that subdivision of the order (*Dasypodidæ*) which is characterized by having perfect and sometimes complex molar teeth, and an external osseous and tesselated coat of mail. The Megatherium is the giant of this tribe; which, at the present day, is exclusively represented by South American species, the largest (*Dasypus Gigas*, Cuv.) not exceeding the size of a Hog. The hiatus between this living species and the Megatherium, is filled up by a series of Armadillo-like animals, indicated more or less satisfactorily by Mr. Darwin's fossils, some of which species were as large as an Ox, others about the size of the American Tapir. The rest of the collection belongs, with the exception of some small Rodents, to the extensive and heterogeneous order Pachydermata; it includes the remains of a Mastodon, of a Horse, and of two large and singular aberrant forms, one of which connects the Pachydermatous with the Ruminant Order; the other, with which the descriptions in the following pages commence, manifests a close affinity to the Rodent Order.

A DESCRIPTION OF THE CRANIUM OF

TOXODON PLATENSIS;

A gigantic extinct mammiferous animal, referrible to the Order Pachydermata, but with affinities to the Rodentia, Edentata, and Herbivorous Cetacea.

THE cranium, which is the subject of the present description, was found in the Sarandis, a small stream entering the Rio Negro, and about 120 miles to the N.W. of Monte Video: it had been originally embedded in a whitish argillaceous earth, and was discovered lying in the bed of the rivulet, after a sudden flood had washed down part of the bank.

The zoological characters deducible from this cranium, forbid its association, generically, with any known Mammiferous animal, and it must therefore be referred to an extinct genus, which I propose to call *Toxodon*,* from the curved or arched form of the teeth, as will afterwards be described. The specific name, in the absence of other means of knowing the peculiarities of the animal than those afforded by the skull, may be most conveniently taken from the district (La Plata), in which its remains were first discovered.

The dimensions of the cranium of the *Toxodon Platensis* amply attest that the animal to which it belonged was of a magnitude attained by few terrestrial quadrupeds, and only to be compared, in this respect, with the larger Pachyderms, or the extinct Megatherium. The length of the skull (of which a base view of the natural size is given in Plate I.) is two feet four inches:

---

* Τοξον, arcus; οδους, dens.

the extreme breadth one foot four inches. The other requisite admeasurements are given in the table at the conclusion of this description.

The general form of the skull, as seen from above, is pyriform; but viewed sideways, and without the lower jaw, it is semi-ovate; it is depressed, elongate, of considerable breadth, including the span of the zygomatic arches, but becoming rather suddenly contracted anterior to them, the facial part thence growing narrower to near the muzzle, which again slightly expands.

Among the first peculiarities which strike the observer, is the aspect of the plane of the occipital foramen, and of the occipital or posterior region of the cranium, the latter of which inclines from below upwards and forwards at an angle of 50° with the basal line of the skull. This slope of the back part of the skull is one of the characteristics of the Dinotherium; it is common to all the Cetacea, and is met with in a slighter degree in many Rodentia, and in the great Ant-eater and some others of the Edentate order. The corresponding aspect of the *foramen magnum* presents nearly the opposite extreme to man in the occipital scale, proposed by Daubenton to determine the diversities of the form of the cranium, as a gage of the intelligence of different animals[†]; and the indication of the limited capacity of the Toxodon, thus afforded, is strengthened by the very small proportion, which the bony walls of the cerebral cavity bear to the zygomatic and maxillary parts of the skull, and to the size of the vertebral column, as indicated by the condyloid processes, and foramen magnum.

The zygomatic arches are of remarkable size and strength; they commence immediately anterior to the sides of the occipital plane, increase in vertical extent as they pass outwards, forwards and downwards, and are suddenly contracted as they bend inwards to abut against the sides of the sockets of the two

---

† Mem. de l'Acad. des Sciences de Paris, 1764, p. 568.

posterior molar teeth.

The cranial cavity is remarkably narrow at the space included by the zygomatic arches; being, as it were, excavated on each side to augment the space for the lodgment of the temporal muscles, so that its diameter at this part is less than that of the anterior extremity of the upper jaw. The upper surface of the cranium expands to form the post-orbital processes, and again contracts anterior to these.

The muscular ridges, or other characters, at the top of the skull, cannot be precisely determined, as a great proportion of the outer table of the bone is broken away, exposing a coarse and thick diplöe. There seems, however, to have been a strong ridge separating the occipital from the coronal or upper surface of the cranium. The form of the remaining parts, which are modified in relation to the attachment of the muscles of the jaws, indicates that these were powerfully developed both for the offices of mastication and prehension. The general form of the skull, while it presents certain points of resemblance to that of the aquatic Pachydermata, and even of the Carnivora, has much that is peculiar to itself; but, in the facial part, it approaches the nearest to that of the Rodentia; and the dentition of the Toxodon, as exhibited in the upper jaw, corresponds with that which characterizes the Rodent Order.

The teeth of the Toxodon consist of molars and incisors, separated by a long diastema, or toothless space. In the upper jaw the molars are *fourteen* in number, there being seven on each side; the incisors *four*, one very large, and one small, in each intermaxillary bone.

The general form and nature of the teeth are indicated by the sockets; and the structure of the grinders is exhibited in a broken molar, the last in the series on the left side of the jaw of the present cranium (See a figure of the grinding surface restored of this tooth, fig. 2, Pl. I.), and by another perfect molar, the last but one on the right side of the upper jaw, which, though not belonging to the same individual as the skull here described, undoubtedly appertains to the same species.

This latter tooth (Fig. 3, Pl. I.; figs. 2 and 3, Pl. IV.) was found by itself, embedded in the banks of the Rio Tercero, or Carcarana, near the Parana, at the distance of a hundred and eighty miles from the locality where the head was discovered. Fragments of a molar tooth of a Toxodon, apparently the seventh of the left side, upper jaw, were also found at Bajada de Sta Fé, in the province of Entre Rios, distant forty miles from the mouth of the Rio Tercero.

All the molar teeth are long and curved, and without fangs[*], as in most of the herbivorous species of the Rodent Order: in those, however, with curved grinders, as the *Aperea* or Guinea-pig, and *Cavia Patachonica*, the concavity of the upper grinders is directed outward, the fangs of the teeth of the opposite sides diverging as they ascend in the sockets; but, in the Toxodon, the convexity of the grinders is outward, and the fangs converge and almost meet at the middle line of the palate, forming a series of arches, capable of overcoming immense resistance from pressure. (See the upper view of the skull, Plate III., in which the fractures expose to view a part of the series of these arched sockets.)

Of the incisors, the two small ones (the sockets of which are indicated at *s s*, Pl. III.) are situated in the middle of the front of the upper jaw, close to the suture between the intermaxillaries, and the two large ones in immediate contiguity with the small incisors, which they greatly exceed in size. The sockets of the two large incisors (*t t*, Pl. III.) extend backwards, in an arched form, preserving a uniform diameter, as far as the commencement of the alveoli of the molar teeth: the curve which they describe is the segment of a circle; the position, form, and extent of the sockets of these incisors are the same as in those of the corresponding teeth of the Rodentia.

The matrix, or secreting pulp of the large incisors, was lodged, as in the Rodentia, in close proximity with the sockets

---

[*] True fangs exist only in teeth of temporary growth, they may be one or more in number, but always diminish in size as they recede from the crown of the tooth, and are either solid, or with a very small canal.

of the anterior molars; and we are enabled to infer, from the form of the incisive sockets, notwithstanding the absence of the teeth themselves, that the pulp was persistent, and that the growth of these incisors, like those of the Rodentia, continued throughout life.

This condition, joined with the form and curvature of the socket, implies a continual wearing away of the crown of the tooth by attrition against opposing incisors of a corresponding structure in the lower jaw: and as a corollary, it may be inferred that the teeth in question had a partial coating of enamel, to produce a cutting edge, and were, in fact, true *dentes scalprarii*. The number of incisors in the upper jaw of Toxodon, is not without its parallel in the Rodent Order, the genus *Lepus* being characterized by four, instead of two superior incisors, which also present a similar relative size but have a different relative position, the small incisors, in the hare and rabbit, being so placed immediately behind the large pair, as to receive the appulse of the single pair of incisors in the lower jaw.

In the Toxodon the position of the incisors, in the same transverse line, might lead to the inference, that they were opposed by a corresponding number in the lower jaw; but the numerous examples of inequality, in the number of incisors, in the upper and lower jaws of existing mammalia, forbid any conclusion on this point.* The sockets of the small mesial incisors of the Toxodon (*s s*, Pl. III.) gradually diminish in size, as they penetrate the intermaxillary bones, and we may, therefore, infer that the pulp was gradually absorbed in the progress of their development; and that, like ordinary incisors, their growth was of limited duration, and their lodgment in the jaw effected by a single conical fang.

I may observe, that the formation of a fang is the necessary

---

* This was written before an examination of the fragment of a lower jaw, forming part of Mr. Darwin's collection of Fossil Remains, had led me to suspect that it was referrible to the genus Toxodon; should this suspicion prove correct, the four unequal incisors of the upper jaw are opposed to six equal sized ones in the lower.

consequence of the gradual absorption of the matrix or pulp of a tooth; for the pulp continues, as it diminishes in size, to deposit ivory upon the inner surface of the cavity of the tooth from which it is receding, and the tooth or fang thus likewise progressively diminishes in size. The formation of the socket proceeds uninterruptedly, and the bone encroaching upon the space left by the tooth, closely surrounds the wasting fang, and affords it a firm support; and thus an inference may be drawn from the form of the socket alone, as to whether the tooth it contained had or had not one or more conical fangs, and consequently whether its growth was temporary or uninterrupted.

Applying this reasoning to the molar teeth of the Toxodon, we infer that their growth, like those of most of the Phytiphagous Rodents, of the Megatherium and Armadillo, was perpetual, because their sockets are continued of uniform size from the open to the closed extremity; and the molar tooth which is preserved proves the accuracy of the deduction, inasmuch as its base is excavated by a large conical cavity for the lodgment of the pulp, the continued activity of which was the compensation here designed to meet the effects of attrition on the opposite or grinding surface of the tooth.

The molar tooth discovered by Mr. Darwin in the banks of the Tercero, not only belonged to the same species as the skull under consideration, but to an individual of the same size; it fits exactly into the socket next to the posterior one of the right side. The figures subjoined of this molar tooth (Fig. 3, Pl. I.; figs. 2 and 3, Pl. IV.) almost preclude the necessity of a description. The transverse section of the tooth gives an irregular, unequal sided, prism; the two broadest sides of which converge to the anterior angle, which is obtusely rounded. The outer surface of the tooth (fig. 2, Pl. IV.) is slightly concave in the transverse direction, but undulating, from the presence of two slight convex risings which traverse the tooth lengthwise. The inner surface presents at its anterior part a slightly concave surface, and posteriorly two prominent longitudinal convex ridges, separated by a groove which is flat at the bottom, and from the anterior angle of which the reflected fold of enamel penetrates the substance of the tooth, advancing obliquely forwards, rather more than halfway across the body of the tooth. A longitudinal ridge of bone projects from the internal side of the socket, and fits into the groove above mentioned, and as a corresponding ridge exists in all the sockets of the grinders, save the two anterior small ones, we may infer that the five posterior grinders on each side, had a similar structure to the tooth above described. The external layer of enamel is uniformly about half a line in thickness; it is interrupted for the extent of nearly three lines at the anterior angle, and for more than double that extent at the posterior part of the tooth, which is consequently worn down much below the level of the rest of the grinding surface. Where the ivory is thus unprotected by the enamel, it has a coat of cæmentum, which also fills up the small interval at the origin of the reflected fold of enamel. On the grinding surface of the entire tooth, and on the fractured ends of the mutilated molars, the component fibres, or tubules, of the ivory, are readily perceptible by the naked eye, diverging from the line which indicates the last remains of the cavity of the pulp of the tooth, as it was progressively obliterated during growth.

Although the complication of the grinding surface by the inflection of simple or straight folds of enamel is peculiarly characteristic of the Rodent type, we must regard the number of molar teeth, and their diminution of size as they advance towards the anterior part of the jaw, in the Toxodon, as indicative of a deviation from that order, and an approach to the Pachyderms. The common number of grinders in the upper jaw of Rodent animals is eight, four on each side. In some genera, as Lemmus, Mus, Cricetus, there are only three on each side, and in Hydromys and Aulacodus, only two on each side. In Lepus, however, we find six on each side of the upper, and five on each side of the lower jaw. The Toxodon, like the Tapir and Hippopotamus, has seven on each side of the upper jaw: the first in each of these species being the smallest. It is worthy of notice, however, that the Capybara which adheres to the Rodent type in the number of its molars, presents in the vastly increased size, and additional number of component laminæ of the posterior grinders, an approximation to the pachydermatous character just adduced, and the bony palate at the same time presents an expansion between these molars, offering a resemblance to the Toxodon which I have not found in any other Rodent besides the Capybara.

The most important deviation from the Rodent structure presented by the teeth, occurs in the direction of the reflected fold of enamel, and such a deviation might have been inferred, even in the absence of the teeth, from the structure of the articular surface, or glenoid cavity for the reception of the condyle of the lower jaw. As the ridge of enamel runs, as above described, in a direction approaching that of the longitudinal axis of the skull, it is obvious that the grinding motions of the lower jaw should be in a proportionate degree in the transverse direction. The glenoid cavity, therefore, instead of being a longitudinal groove, and open behind, as in the true Rodents, is extended transversely, and is defended behind by a broad descending bony process preventing the retraction of the jaw, and showing marks of the forcible pressure to which it was subject.

It is worthy of observation that, in the Wombat,—which exhibits the Rodent type of dentition, and, like the Toxodon, has remarkably curved molars, but in an opposite direction,— the condyle of the lower jaw is also extended transversely, and adapted to an articular surface, which admits of lateral motion in the trituration of the food. In the outward span of the zygomatic arches, in which Toxodon deviates from the Rodentia, we may trace a relation of subordinacy to the above structure of the grinding teeth and joint of the lower jaw: the widening of the arches giving to the masseter muscles greater power of drawing the jaw from side to side. The depth of the zygoma bespeaks the magnitude of these masticatory muscles, and the included space shews that the temporal muscles were also developed to a degree, which indicates the force with which the great incisors at the extremity of the jaws, were used; probably, like the canines of the Hippopotamus, to divide or tear up by the roots the aquatic plants, growing on the banks of the streams, which the Toxodon may have frequented.

In the Rodentia, the zygoma, though sometimes as deep as in the Toxodon, is generally almost straight, and the space included between it and the cranium is consequently narrow. The zygoma also is placed more forwards in all true Rodents, than in the Toxodon; and, instead of abutting against the posterior alveoli, it terminates opposite the anterior ones. It thus affords such an attachment to the masseter, that this muscle extends obliquely backwards to its insertion in the lower jaw, at an angle which enables it to act with more advantage in drawing forwards the lower jaw,—a motion for which the joint is expressly adapted. In many Rodents, also, there is a distinct muscle, or portion of the masseter, which passes through the ant-orbital foramen, which is on that account of large size. In examining the cranium of Toxodon, with reference to this structure, it was found that the ant-orbital foramen was not larger than might have been expected to give transmission to nerves requisite for supplying with sensibility the large lips, and whiskers with which the expanded muzzle of this remarkable quadruped was probably furnished.

Having thus examined the cranium of the Toxodon in its relation, as a mechanical instrument, subservient to the function of digestion; we next proceed to consider the structure and composition of those cavities of the skull which gave lodgment and protection to the organs of *special* sense, and endeavour to deduce from their structure conclusions as to the degree in which the organs were developed, and the circumstances under which the senses were exercised.

The orbit of Toxodon forms the anterior boundary of the zygomatic area; it is about as distinctly defined as in the Tapir or Dugong, having its osseous rim less complete than in the

Hippopotamus, yet more developed than in the Capybara, Coypus, and many other Rodentia, in which the orbit is scarcely distinguishable in the cranium from the small space occupied by the origin of the temporal muscle.

The lower boundary of the orbit in Toxodon is formed by an excavation in the upper and anterior part of the zygoma; the upper boundary by a strong and rugged overarching process of the frontal bone, the posterior angle of which (*a*, Pl. III.) descends a little way, but leaves a space of three inches and a half between it and the opposite angle of the malar bone below (*b*, Pl. II. and III.), the circumference of the orbit being completed probably by ligament in the recent subject. The cavity thus circumscribed is remarkable for the preponderance of the vertical over the transverse or longitudinal diameter, and indicates great extent of motion of the eyeball in the vertical direction, such as may be supposed to be well adapted to the exigencies of an amphibious quadruped. The orbit of the Capybara, or Water-hog, makes a near approach to the form just described. In the elevation of the supra-orbital boundary, and its outward projection in the Toxodon, we perceive an approximation to the form of the orbit in the Hippopotamus, but the size of the orbit is relatively larger in the Toxodon, which in this respect manifests its affinity to the Rodentia.

In that part of the bony structure of the auditory apparatus, which is visible on the exterior of the cranium, the skull of the Toxodon presents a character in which it recedes from the Rodentia. In these, the tympanic portion of the temporal bone is remarkably developed, forming a large bulla ossea between the glenoid cavity and the occiput; and it always remains disunited to the other elements of the temporal bone. In the Toxodon the tympanic bone (*c*, Pl. II.) consists of a rough compressed vertical osseous plate, wedged in transversely between the occiput and the posterior part of the glenoid cavity. The internal extremity of this plate points inwards and forwards, representing the styloid process; behind this is seen the petrous bone, which forms a small angular protuberance at the basis cranii, and is less developed than in the Hippopotamus. Anterior to the petrous bone are the orifices of the Eustachian tube, and carotid canal; external to it is the great foramen lacerum, for the jugular vein and nervus vagus; and behind it is the anterior condyloid foramen. The foramen auditorium externum is only half an inch in diameter, and gives passage to a long and somewhat tortuous meatus, which passes inwards and slightly forwards and downwards; its direction being precisely the same as in the Hippopotamus; it was accompanied, probably, by as small an external auricle.

But the indications of the aquatic habits of the Toxodon, which are presented by the osseous parts relating to the senses of sight and hearing, are of minor import compared with those afforded by the bony boundary of the nostrils. This boundary circumscribes a large ovate aperture, the aspect of whose plane is upwards, and a little forwards, as in the Herbivorous Cetaceans, and especially the Manatee (*Trichecus Manatus*, Cuv.) In one part of the bony structure of the nasal cavity the Toxodon deviates, however, in a marked degree from the Cetaceous structure; I allude to the frontal sinuses, which are exposed by the fracture of the upper part of the skull. (They are shewn in Plate III., and an asterisk is placed on one of the narrow canals of intercommunication between the sinuses and the nasal passages.) The posterior orifice of the nasal cavity is relatively larger and wider than in the Herbivorous Cetaceans, and differs both in form and aspect in consequence of the greater extent of the bony palate. The Toxodon further differs from the Manatee and Dugong, in the firm nature of the connexion of the bones of the head; and it differs from the Hippopotamus in the strong attachment of the intermaxillary bones to the maxillaries.

There next remain to be described, as far as the shattered condition of the skull will permit, the relative position, extent, and connexions of the principal bones composing it.

The *occipital bone* exhibits a complete confluence of its basilar, condyloid, and supra-occipital elements. The basilar portion, in connexion with the corresponding element of the sphenoid bone, describes a curve whose convexity is downwards. The condyles are large, extended in the transverse direction, completely terminal, and a little inclined downwards below the level of the basilar process. The curve of the articulating surface describes, in the vertical direction, two-thirds of a circle, indicating that the head must have possessed considerable extent of motion upwards and downwards upon the atlas; thus, while the body of the Toxodon was submerged, the head probably could be raised so as to form an angle with the neck, and bring the snout to the surface of the water without the necessity of any corresponding inflection of the spine. Indeed, in the form and position of the condyles, the Toxodon more nearly resembles the true Cetacea than any other existing mammalia; and it is only with these that it can be compared in regard to the aspect of the plane of the occipital foramen, and of the occipital region of the skull. This is inclined forwards from the occipital foramen at such an angle, that on viewing the skull from above, not only the condyles, but the entire circumference of the occipital foramen are visible. (See Pl. III.) The upper part of the supra-occipital plate presents a broad rugous depression, indicative of the insertion of strong cervical muscles, and probably of a *ligamentum Nuchæ*.*

The ex-occipital processes advance forwards for about an inch beyond the condyles, and then suddenly extend outwards at right angles to the former line, and terminate in the form of vertically compressed bony plates; the lower rugged margins of which represent or perform the office of the mastoid processes (*d, d*, Pls. II. and III.). The breadth of the entire occipital region of the skull (fig. 1, Pl. IV.) appears to have been, allowing for the fractures, about one-third more than the height of the same part.

The great development of the *tympanic* bones in the Rodentia, occasions the intervention of a considerable space between the occipital bone and the zygomatic process of the temporal; but in

---

* I have ascertained that this elastic ligament exists in the neck of the Dugong.

the great Toxodon, in which the sense of hearing was doubtless inferior to that enjoyed by the small and timorous Rodents, the tympanic bone is reduced to a thin plate, which is wedged in between the occiput and glenoid cavity. In this structure, and the consequent posterior position of the glenoid cavity, there is a close resemblance between the Toxodon and the Hippopotamus, Tapir, and Rhinoceros.

The *squamous* element of the temporal bone (N, Pl. II.) forms a small proportion of the lateral walls of the cranium, and also enters into the composition of the lateral and superior parts of the posterior region of the cranium, where two deep fossæ perforated by large vascular foramina, indicate the junction of the squamous bones with the supra-occipital bone. The posterior surface of the skull is thus divided into three broad and shallow depressions, the two lateral facets being slightly over-lapped by the middle one, at their junction with it. In this structure the Toxodon rsembles the Hippopotamus, and differs considerably from the Cetacea, in which the occipital region is rendered convex by the extraordinary development of the brain within.

The *zygomatic* process of the temporal bone projects boldly outwards at its commencement, where it is of great strength, and three-sided; the glenoid cavity extends transversely across the base or inferior surface of this part; the lateral surfaces converge to form the ridge or upper boundary of the zygoma. The depth of the glenoid cavity is increased by a transverse production of bone both before and behind it: the posterior process (*g*, Pl. II.) descends the lowest, and affords the requisite defence against backward dislocation of the lower jaw; the pressure of the condyle against this process is denoted by a well defined, transversely-ovate, flattened and smooth surface, as if the bone had been planed down at that part: the anterior transverse boundary is convex and smooth, and probably formed part of the articulation for the lower jaw. The lower facet of the zygoma anterior to the glenoid cavity gradually contracts in breadth, as it advances forward, and at the distance of three inches from the articular cavity the zygoma changes from a prismatic to a laminar form. It is at this point that the zygomatic suture commences, at the lower margin of the arch; whence it extends directly forwards for more than half its length, and then bends upwards at a right angle. The zygomatic suture has a similar course in the Capybara, and Hippopotamus.

The remainder of the zygoma is formed externally by the *malar* bone (G Pl. II.), which in its position is intermediate to the Rodent and Pachydermatous structures. It is not suspended in the middle of the zygomatic arch, as in the former order; neither does it extend into the region of the face so far anterior to the orbit as in the Tapir or Hippopotamus. The exterior line of the malo-maxillary suture defines the orbit anteriorly; but from this line the maxillary bone extends backwards, along the inner side of the malar portion of the zygoma, until it almost reaches the temporo-malar suture; thus abutting by an oblique surface against nearly the whole internal facet of the malar bone, and materially contributing to the general strength of the zygomatic arch. The malar bone is of considerable vertical extent, and presents a rugged and thickened inferior margin for the attachment of the masseter. The upper margin of the malar bone is smoothly rounded, and presents a regular semi-circular excavation, forming the lower boundary of the orbit. The relative magnitude of the zygomata to the entire cranium far exceeds in the Toxodon that which exists in the Hippopotamus or any other known Pachyderm. This arises from the great vertical development of the malar bone behind the orbit, and the vertical expansion of the temporal portion of the arch. The oblique position of the zygoma, descending as it advances forwards, is deserving of attention, as the Toxodon, in deviating from the Pachyderms in these respects, makes an evident approach to the herbivorous Cetaceans, as the Dugong and Manatee: in the latter Cetacean we observe a similar development of the lower part of the zygomatic process of the malar bone. It is here, also, that we may perceive an indication of a resemblance between the

Megatherium and Toxodon.

There is no discernible trace of the *lachrymal* bone (E, Pl. II.) having extended, as in the Hippopotamus beyond the anterior boundary of the orbit: the lachrymal foramen is situated rather deep in the orbit, and the bone itself appears to have been of very small size.

The surface of the supra-orbital process of the *frontal* bone (C, Pl. II.) is deserving of attention, as it presents a peculiar ruggedness which is not found in any other part of the skull; the irregularity seems, as it were, to have been produced by the impression of numerous small tortuous and anastomosing vessels. In the skull of a Sumatran two-horned Rhinoceros, in the Museum of the College of Surgeons (No. 816), the circumference of that part of the surface of the skull which supported the posterior horn, and which includes precisely the same part of the os frontis, presents the same character, the surface being broken by numerous vascular impressions. On the supposition that this character of the supra-orbitary arch in the Toxodon might indicate the superincumbency of a bony case, I examined the skulls of two Armadillos, *Dasypus Peba* and *Das. 6-cinctus*, and found that in the Dasypus 6-cinctus, the supra-orbital ridges, which are slightly elevated, to support the cephalic plate, presented, in a minor degree, a corresponding rugosity. May we venture then to conjecture that the Toxodon was defended by an ossified integument like the Armadillo, or that it was armed with an epidermic production, analogous to the horn of the Rhinoceros; or had the rugous surface in question as little relation with the parts that covered it as the sculptured surface of the malar bones in the Cavy?

After forming the rugged and prominent supra-orbital processes already described, the frontal bone continues to send backwards a slightly elevated ridge or *crista*, circumscribing the origin of the temporal muscles, but the extent of this ridge, and the disposition of the inter-orbital portion of the frontal bones cannot be determined in the present mutilated specimen. The fractures it has sustained are not, however, wholly unattended with advantage; they expose the structure of the diploë, which from its coarseness of texture and thickness, resembles that of the Cetaceous crania; and what is of still more importance, they also demonstrate the existence and form of the frontal sinuses.

The cavity of the nose is extensive, and the remains of the ossa spongiosa superiora testify that the Toxodon enjoyed the sense of smell to a degree equal at least to that of the Hippopotamus.

The *sphenoid bone* resembles that of the Hippopotamus, but it contributes a larger share to the formation of the internal pterygoid processes (p, Pl. II.); these are of a simple form, and more developed than in the Hippopotamus; they project outwards to a greater extent, and terminate in a point. The sphenoid also sends off a short and thick pointed process from the posterior part of the base of the internal pterygoid processes. The ala of the sphenoid does not rise so far into the orbit, nor does it articulate with the parietal bone, as in the *Hippopotamus*; but in this part of its structure, is the same as in the Rhinoceros. The spheno-palatine foramen is relatively larger than in the above-named Pachyderms, and is bounded above by the descending orbital plate of the frontal bone.

The palatal processes of the *palatine* bones terminate anteriorly between the last molars, and extend backwards for some distance beyond the alveolar processes, increasing the extent of the bony roof of the mouth posteriorly: this is a structure in which the Toxodon deviates both from the Rodents, and Pachyderms, and resembles the Armadillos among the Edentata; excepting that the post-dental part of the bony palate in the Toxodon is suddenly contracted in breadth. The palato-maxillary suture is in the form of a chevron, with the angle directed forwards, as in the Hippopotamus and Capybara, but truncated.

The *superior maxillary* bones (F, Pl. II.) are united posteriorly to the malar, as above described: they ascend and join the frontal and nasal bones: their outer surface is almost vertical, smooth, and slightly undulating; perforated at its posterior part by the ant-orbital foramen, and joined anteriorly to the intermaxillaries by a suture running in the sigmoid direction (as shewn in Pl. II.) from the middle of the nasal cavity, to within four inches of the anterior boundary of the upper jaw. We have, in the position and extent of this suture, and the absence of tusks and their large prominent sockets, a most important difference between the Toxodon and Hippopotamus. The chief peculiarity in the maxillary bones, obtains in the arched form of the alveolar processes, corresponding to the shape and position of the grinders above described, and which are peculiar among known mammalia to the present genus. The palatal surface of the maxillary bones is obliquely perforated by two large foramina, from which two deep longitudinal grooves extend forwards, and are gradually lost; we find the posterior palatine foramina represented by similar grooves and foramina in the Capybara.

The *intermaxillary* bones (D, Pls. II. and III.), though large, are relatively of less extent than in the Rodents generally. The nasal processes do not reach the frontal bone, but are limited to the anterior half of the nasal boundary; approaching in this respect to the Herbivorous Cetacea. In the outward expansion of their anterior extremities, the intermaxillaries resemble those of the Hippopotamus, in which, however, this character is more strongly marked. The intermaxillaries in the Hippopotamus are also much less firmly united to the maxillary bones than in the Toxodon, and are consequently commonly lost in the fossil crania. On the palatal surface of the intermaxillary bones there are two grooves which diverge forwards from the line of the suture; and anteriorly to these grooves there are the two large anterior palatine foramina. The maxillo-intermaxillary sutures on the palate converge as they extend backwards to a point; there appears to have been a fissure left between this suture and the mesial suture of the intermaxillaries; in which structure the Toxodon resembles the Hippopotamus.

After summing up the different affinities, or indications of affinity, which are deducible from the cranium of this most curious and interesting fossil mammal, we are led to the conclusion, assuming it to have had extremities cased in hoofs, that it is referrible to the Order Pachydermata. But the structure, form, and kind of teeth in the upper jaw, prove, indisputably, that the gigantic Toxodon was intimately related to the Rodent Order. From the characters of this order, as afforded by the existing species, the Toxodon, however, differs in the relative position of the supernumerary incisors, and in the number, and direction of the curvature, of the molars. If, moreover, the lower jaw, next to be described, belong, as I believe, to the Toxodon, the dental character of the genus will be *incisors 4/6; pro laniariis diastema; molares 7/7 7/7.*

The Toxodon again deviates from the true Rodentia, and resembles the Wombat, and the Pachyderms, in the transverse direction of the articular cavity of the lower jaw.

It deviates from the Rodentia, and resembles the Pachydermata in the relative position of the glenoid cavities and zygomatic arches, and in many minor details already alluded to.

In the aspect of the plane of the occipital foramen, and occipital region of the skull; in the form and position of the occipital condyles; in the aspect of the plane of the anterior bony aperture of the nostrils; and in the thickness and texture of the osseous parietes of the skull, the Toxodon deviates both from the Rodentia and existing Pachydermata, and manifests an affinity to the Dinotherium and Cetaceous Order, especially the Herbivorous section.

At present we possess no evidence to determine whether the extremities of the Toxodon were organized on the ungulate or unguiculate type, nor can we be positive, from the characters which the skull affords, that the genus may not be referrible to the *Mutica* of Linnæus;* although the development of the nasal

---

* The German Translator (See *Frorieps Notizen.*, 1837, p. 119) of the abstract of my description of the Toxodon, published in the Proceedings of the Geological Society, asks, what is the *Mutica* (misprinted Muticata), of Linnæus? The term is quoted from the Systema Naturæ, Ed. xii. p. 24. Linnæus first divides Mammalia

cavity and the presence of large frontal sinuses render it extremely improbable that the habits of this species were so strictly aquatic, as the total absence of hinder extremities would occasion.

Where the dentition of a mammiferous animal is strictly carnivorous, this structure is obviously incompatible with a foot incased in a hoof:—but where the teeth are adapted for triturating vegetable substances the case is different. If animals so characterized are of small size and seek their food in trees, or if they burrow for roots or for shelter, the vegetable type of dentition must co-exist with unguiculate extremities, as in the Edentata and Rodentia generally: but the largest genus (Hydrochærus) of the Rodent Order, whose affinity to the Pachydermata is manifested in its heavy shapeless trunk, thinly scattered bristly hair, and many other particulars, has each of its toes inclosed in a miniature hoof.

The affinity above alluded to, is too obvious to have escaped popular notice, and the Capybara, from its aquatic habits, has obtained the name of Water-hog. It is highly interesting to find that the continent to which this existing aberrant form of Rodent is peculiar, should be found to contain the remains of an extinct genus, characterized by a dentition which closely resembles the Rodent type, but manifesting it on a gigantic scale, and tending to complete the chain of affinities which links the Pachydermatous with the Rodent and Cetaceous Orders.

---

into three groups, according to modifications of the locomotive organs, viz. *Unguiculata, Ungulata, Mutica*, and subdivides these, according to modifications of the dentary organs, into the orders, *Bruat, Glires, Primates*, &c.

ADMEASUREMENTS OF THE CRANIUM OF TOXODON.

	feet	inches	lines
Extreme length...........	2	4	...
Extreme breadth...........	1	4	..
Extreme height, (exclusive of the lower jaw).....	...	10	...
Length of zygomatic process.........	1	1	6
Depth or vertical extent of do.........	...	6	...
Transverse extent of zygomatic fossa........	...	6	...
Transverse diameter of cranium between the zygomatic arches...	...	5	...
Transverse diameter of occipital plane of the cranium.....	1	...	...
From the outside of one condyle to that of the opposite condyle...	...	8	6
Length of the bony palate........	1	6	...
Extreme breadth of ditto.........	...	6	...
Breadth of palate at the intermaxillary suture.......	...	2	6
Do. do. behind the molar alveoli......	...	3	...
Longitudinal extent of the molar alveoli.....	...	9	6
Do. do. diastema...........	...	5	6
Transverse diameter of posterior nasal aperture.....	...	3	9
Do. do. of occipital foramen......	...	3	...
Do. do. of glenoid cavity..........	...	4	6
Antero-posterior do of ditto............	...	1	...

DESCRIPTION OF FRAGMENTS OF A LOWER JAW AND TEETH OF A TOXODON.

Found at Bahia Blanca, in latitude 39° on the East coast of South America.

IN looking over some fragments of jaws and teeth, forming

part of Mr. Darwin's collection of South American mammiferous remains, and which had been set aside with mutilated specimens referrible to species belonging to the family of Edentata, my attention was caught by the appearance of roots of teeth projecting, in a different direction from the grinders, from the fractured anterior extremity of a lower jaw, and I was induced to examine minutely the structure of the teeth in this specimen, and to search the collection for corresponding fragments. The result was the discovery of portions of the two rami, and the commencement of the symphysis of a lower jaw, containing anteriorly the roots of six incisors, and at least six molars on each side; but as the rami had been fractured through the middle of the sixth alveolus, the number of grinders may have corresponded with those in the upper jaw of the Toxodon.

The most perfect of these fragments is figured in Pl. V. figures 1 and 4; figure 2 shows the form of the teeth in transverse section, and the disposition of the enamel upon the grinding surface of the molars on the right side, as restored from a comparison of the fractured teeth in the two rami. From the remains of the symphysis shown at fig. 4, it will be seen that the jaw was remarkably compressed, or narrow from side to side; while the rami (fig. 1.) were of considerable depth, in order to give lodgment to the matrices and bases of grinders enjoying uninterrupted growth.

The pulps of the six incisors in this lower jaw are arranged in a pretty regular semi-circle, whose convexity is downwards; the teeth themselves are directed forwards, and curved upwards, like the inferior incisors of the Rodentia. The form and degree of the curvature are shown in the almost perfect incisor (Pl. V. fig. 5) which corresponds with the left inferior incisor of the lower jaw, and was found in the same stratum, but belonged to another individual.

These incisors are nearly equal in size: they are all hollow at their base, and the indurated mineral substance impacted in their basal cavities well exhibits the form of the vascular pulps which formerly occupied them. Sufficient of the tooth itself remains in four of the sockets to show that these incisors, like the nearly perfect one (fig. 5), had only a partial investment of enamel; but though in this respect, as well as in their curvature and perpetual growth, they resemble the dentés scalprarii of the Rodentia, they differ in having a prismatic figure, like the inferior incisors of the Sumatran Rhinoceros, or the tusks of the Boar. Two of the sides, viz., those forming the anterior convex and mesial surfaces of the incisor have a coating of enamel, about half a line in thickness, which terminates at the angles between these and the posterior or concave surface. In plate V. fig. 4, the enamel of the broken incisors is represented by short lines, showing the direction of its crystalline fibres; the white space immediately within the enamel shows the thickness of the ivory at the base of the tooth, the included gray substance represents a section of the formative matrix or pulp of the tooth, which was of the usual conical form: the inferior broken end of the incisor (fig. 5,) appears to have been distant about one-third from the apex of the pulp.

From the relative position of the bases or roots of these incisors, we may infer that they diverged from each other as they advanced forwards, in order to bring their broadest cutting surfaces into line. That they were opposed to teeth of a corresponding structure in the upper jaw is proved by the oblique chisel-like cutting surface of the more perfect incisor: and it is not without interest to find that the presence of *dentes scalprarii* at the anterior part of the mouth has not been necessarily limited to Mammalia of small size.

The position of the pulps of these incisors, in close proximity with the anterior grinders, corresponds with the position of the pulps of the incisors in the upper jaw of the *Toxodon*, and indicates, in conjunction with the size of the pulps, that a considerable extent of the inferior incisors was lodged in the substance of the anterior part of the jaw. It is most likely that no vertically directed tooth would be developed in the part of the jaw so occupied by the curved bases of the incisors, and hence a diastema or toothless space would intervene between the molars and incisors of this lower jaw, as in the upper jaw of the Toxodon.

It is interesting, also, to observe, that as the deviations from the Rodent type, which occur in the cranium of the Toxodon, are the same, in some instances, as those which obtain in the Wombat; so we find a corresponding deviation in the size and relative position of the inferior incisors, which, as in the Wombat, terminate anterior to the molar teeth, instead of extending backwards beyond the last grinder, as in most of the true Rodents. The Capybara presents the nearest approach to this structure, the pulps of the inferior incisors being situated opposite the interspace of the first and second grinders.

The molar teeth, in this mutilated lower jaw, like those in the upper jaw of Toxodon, had persistent pulps, as is proved by the conical cavity at their base, as represented in fig. 3; they consequently required a deep socket, and a corresponding depth of jaw to form the socket and protect the pulps. In order to economise space, and to increase the power of resistance in the tooth, and perhaps, also, to diminish the effects of direct pressure on the highly vascular and sensible matrix, we find the molars and their sockets are curved, but in a less degree than those of the upper jaw of the Toxodon. They correspond, however, with the superior molars of the Toxodon in the antero-posterior diameter, in being small and simple at the anterior part of the jaw, and by increasing in magnitude and complexity as they are situated more posteriorly. They are, however, narrower from side to side; but supposing them to belong to the Toxodon, it would agree in this respect with most other large herbivorous mammalia;—the fixed surface for attrition in the upper jaw being from obvious principles more extensive than the opposed moveable surface in the lower jaw.

The *first* grinder, in the lower jaw here described (Pl. V. fig. 2), is of small size and simple structure, being surrounded with a coating of enamel of uniform thickness, and without any fold penetrating the substance of the tooth. It is more curved than any of the other molars, and appears to have differed from the external incisor only in its entire coating of enamel and direction of growth; it is interesting, indeed, to find so gradual a transition, in structure, from molar to incisive teeth as this jaw presents; for the robust incisors may here be regarded as representing molars simplified by the partial loss of enamel, and with a change in their direction.

In the *second* molar, we find an increase in the antero-posterior diameter, and in the length of the tooth, and the enamel at the middle of the outer side makes a fold which penetrates a little way into the tooth; the line of enamel, on the inner side, is slightly concave and unbroken.

The *third* molar presents an increase of dimensions in the same directions as the second; the enamel on the outer side of the tooth presents a similar fold, but it is directed a little more backwards.

In the *fourth* molar, besides a further increase of size, and a corresponding but deeper fold of enamel in the external side of the tooth, we have the grinding surface rendered more complicated by two folds of enamel entering the substance of the tooth from the inner side: these folds divide the antero-posterior extent of the tooth into three nearly equal parts; they are both directed obliquely forwards, half-way across the substance of the ivory.

The *fifth* molar presents the same structure as the fourth, which it exceeds only slightly in size.

In the *sixth* molar we have a proportionally greater increase of size in the antero-posterior diameter, which measures two inches; but the lateral diameter is but slightly augmented; its structure resembles that of the fifth.

As these grinding teeth by no means increase in the lateral diameter in the same proportion as in their antero-posterior diameter, the posterior ones present, but in a greater degree, the compressed form which characterizes the grinders of the upper jaw of the Toxodon.

It will be seen, however, that there is a difference in the structure of the grinders in this fragment of the lower jaw and those of the upper jaw of the Toxodon. In the lower grinders there are two folds of enamel proceeding from the inner side of the tooth into its substance, whilst in the upper grinders there is only one fold continued from the inner side; in the lower grinders there is also a fold of enamel reflected into the substance of the tooth from the outer surface, while in the upper grinders of Toxodon we find the enamel coating on the outer side of the tooth merely bent inwards, so as to describe, in the transverse section, a gently undulating line; fig. 7, Pl. V. is the grinding surface of the sixth molar, right side, upper jaw.

But this difference of structure is by no means incompatible with the co-existence of the two series of teeth in the same animal, since we find the grinders of the upper and lower jaws presenting differences of structure of equal degree in existing herbivorous species. If we examine the jaws of the Horse, for example, we shall find not only an equal amount of difference in the structure of the upper and lower grinders, but that they deviate from one another in a very similar manner to that above described in the Toxodon. In this comparison attention should be confined to the course of the external enveloping layer of enamel, leaving out of consideration the central crescentic islands of enamel which constitute the additional complexity of the Horse's grinder. Viewing then the course of the external coat of enamel on the worn surface of the tooth, we find it describing on the outer side of the tooth in the upper jaw an undulating line,—a middle convexity being situated between two concavities; on the inner side of the tooth one fold of enamel penetrates to the middle of the tooth, and on each side of this there is a smaller fold. But in the lower jaw the line of enamel on the outer side of the tooth, instead of merely bending outwards midway in its course, is reflected a little way inwards; while on the opposite, or inner side of the tooth, the enamel sends two extensive folds into the substance of the tooth, opposite to the interspace of which the shorter fold projects from the outer side. Now, on the supposition that the fragment of the lower jaw here described belongs to the Toxodon, the kind and degree of difference in the complexity of the grinding surface of the teeth in the upper and lower jaw, are remarkably analogous to those which exist in the Horse. I have only further to remark that in the Horse the inflected folds of enamel, instead of being simple and straight with the two constitutive layers in apposition, as in the Toxodon, are irregular in their course, with cœmentum intervening between the constitutive layers, which also diverge from each other at their angle of reflection, so as to augment the amount of dense material which enters into the composition of the tooth.

Many analogous examples will readily occur to the experienced comparative anatomist. The Horse has been adduced as one to which reference can very readily be made; but I would also cite the Sumatran Rhinoceros, the skull of which, in the Hunterian collection, has already been alluded to. In this species the anterior grinders, in both jaws, are small and simple, and increase in complexity as they recede backwards. The third superior grinder (fig. 8, Pl. V.) presents a single fold of enamel, reflected obliquely forwards from the inner side half-way across the tooth; the outer line of enamel describes a simply undulating line. The opposite grinder of the lower jaw (fig. 9, Pl. V.) has only one-half the breadth of the upper one, but has its grinding surface further complicated by having two inflected folds of enamel from the inner side, and one shorter and broader fold from the outer side. This tooth, therefore, presents a close resemblance to one of the posterior grinders of the lower jaw of the Toxodon, but differs essentially in being of limited growth, and consequently in having fangs.*

In speculating upon the nature of the organized substances

---

* Besides the relation to *food requiring much comminution*, which teeth with persistent pulps bear, they are also connected with the *longevity of the individual*. The term of life in a herbivorous animal, with grinders of temporary growth, is, of necessity, dependent on the duration of these essential aids to nutrition; thus, a sheep generally wears down its grinders in twelve years, and its natural term of life is consequently limited to about that period.

which the teeth of the Toxodon were destined to grind down, we must not only take the structure of the tooth into consideration, but also the power of perpetual renovation, which will compensate for the defective quantity of enamel in the grinders of the Toxodon, as compared with those of the existing Ruminants and Pachyderms, whose grinders, when once completed, receive no further addition of dental substance at their base. The Toxodon, in this character of its dentition, participated in the same advantages with the Capybara and the Megatherium.

Although we have been enabled to observe the structure of the grinding teeth of the upper jaw of the Toxodon in two examples only; one, an insulated perfect grinder corresponding to the sixth alveolus on the right side, and the other, a portion of the last grinder of the left side remaining in the socket of the head previously described, yet from the relations subsisting between socket and tooth, a very satisfactory opinion may be formed of the structure of those teeth which are wanting, as well as of their size. It thus appears, that the grinders of the upper jaw of the Toxodon, are small and simple at the anterior part of the jaw, and that they increase (chiefly in antero-posterior extent) in size, as well as in complexity, as they recede backwards in the jaw. In this respect, as well as in size, the teeth, in the fragments of the lower jaw just described, exactly correspond. There is, however, a slight difference in the lateral diameter of the two sets of grinders, those of the lower jaw being narrower, as is usually the case, but not in the same degree as in the Horse or Ruminant. A greater difference obtains in the degree of curvature of the two sets of molars, those of the lower jaw, especially the posterior grinders, being much less bent than the corresponding teeth of the upper jaw. It is necessary to observe, also, that the convexity of the curve of the inferior grinders is directed outwards, as in the superior grinders; while in the Guinea Pig and Wombat, which have also curved grinders, the convexity is outwards in the lower jaw, and inwards in the upper jaw.

Nevertheless, if we take into consideration the close similarity which exists between the teeth of the upper jaw of the Toxodon, and those of this lower jaw in more essential points, as in their persistent pulps, their characteristic structure and form, the depth of their sockets, and their relative sizes and complexity; and when we consider how the depth of this lower jaw, and its narrowness in the transverse direction, corresponds with the characteristic form of the upper jaw of the Toxodon, and that to these resemblances is added an apparatus of incisors adequate to oppose the great dentes scalprarii of the upper jaw, the conclusion seems irresistiable, that the lower jaw, here described, must be referred, if not to the same, at least to a nearly allied species of Toxodon, as that to which the large cranium belonged.

Further researches in South America, it is hoped, will lead, ere long, to the completion of our knowledge of the osteology of this very remarkable and interesting genus of extinct mammiferous animals.

DESCRIPTION OF PARTS OF THE SKELETON OF

MACRAUCHENIA PATACHONICA;

A large extinct Mammiferous Animal, referrible to the Order Pachydermata; but with affinities to the Ruminantia, and especially to the Camelidæ.

IN the preceding pages the nature and affinities of a large extinct Mammal were attempted to be determined from the cranium and teeth exclusively: we come now to consider the remains of a quadruped consisting of bones of the trunk and extremities, without a fragment of a tooth or of the cranium to serve as a guide to its position in the zoological scale.

It may appear, even to anatomists and naturalists familiar with the kind of evidence afforded by a fossil fragment, that an opinion as to the relation of the present species to a particular family of Ruminants, formed without a knowledge of the important organs of manducation, must be vague and doubtful, but the evidence about to be adduced, will be regarded, it is hoped, as more conclusive than could have been *à priori* expected.

The portions of the skeleton of the animal—which, in relation to the affinity above alluded to, as well as from the length of its neck, I propose to call *Macrauchenia**—were discovered by Mr. Darwin in an irregular bed of sandy soil, overlying a horizontal accumulation of gravel on the south side of Port St. Julian: and independently of the circumstances under which they were found, their correspondence with each other in size, colour, texture and general character prove them to have belonged to one and the same individual.

These remains include two cervical vertebræ, seven lumbar vertebræ, all more or less fractured; a portion of the sacrum and ossa innominata; fragments of the left scapula; of the left radius and ulna, and left fore-foot; the left femur nearly entire, the proximal and distal extremities of the left tibia and fibula; and a metatarsal bone of the left hind foot.

Before entering upon the description of these remains, a few observations may be advantageously premised on some of the distinguishing characters of the Camelidæ. It is well known that the Camels and Llamas deviate in their dentition, viz., in the presence of two incisors in the upper jaw, from the true Ruminants; and we cannot avoid perceiving that in this particular the direction in which they deviate tends towards the conterminous Ungulate Order, in which incisor teeth are

---

* Μακρος *longus,* αυχην *cervix:* from the latter word Illiger derived *Auchenia,* his generic name of the Llama, Vicugna, &c.

rarely absent in the upper jaw. They also further deviate from the Ruminants and approach the Pachyderms in the absence of cotyledons in the uterus and fetal membranes; having, instead thereof, a diffused vascular villosity of the chorion, as in the sow and mare.

But besides these characters, by which, in receding from one type of hoofed mammalia, the Camelidæ claim affinity with another, there are many parts of their organization peculiar to themselves; of some of these peculiarities, the relation to the circumstances under which the animal exists, can be satisfactorily traced; in others, the connection of the structure with the exigencies of the species, is by no means obvious, and in this predicament stands the osteological peculiarity, which is immediately connected with our present subject—a peculiarity in which the Camelidæ differ not only from the other Ruminants, but from all other existing Mammalia, and which consists in the absence of perforations for the vertebral arteries in the transverse processes of the cervical vertebræ, the altas excepted.

I may observe that what is described as a perforation of a single transverse process in a cervical vertebra is essentially a space intervening between two transverse processes, a rudimental rib, and the body of the vertebra. In the cold-blooded Saurians,— in which the confluence of the separate elements of a vertebra takes place tardily and imperfectly, if at all,—the nature of the so called perforation of the transverse process is very clearly manifested, as in the cervical vertebræ of the Crocodile, in which the interspace of the inferior and superior transverse processes is closed externally by a separate short moveable cervical rib. In the Ornithorhynchus paradoxus the vertebra dentata also preserves throughout life this condition of its lateral appendages: in other Mammalia it is only in the fœtal state that the two transverse processes are manifested on each side with their extremities united by a distinct cartilage, which afterwards becomes ossified and anchylosed to them.

In the Hippopotamus the inferior transverse process sends

downwards a broad flat plate extended nearly in the axis of the neck, but so obliquely, that the posterior margins of these processes, in one vertebra, overlap the anterior ones of the succeeding vertebra below, like the cervical ribs in the Crocodile; the same structure obtains in many other mammalia, especially in the Marsupials. In the Giraffe, the inferior transverse processes are represented by relatively smaller compressed laminæ, projecting obliquely downwards and outwards from the anterior and inferior extremity of the body of the vertebra. The superior transverse processes in this animal are very slightly developed in any of the cervical vertebræ, and the perforation for the vertebral artery is above and generally in front of the rudiment of this process, being continued as it were through the side of the substance of the body of the vertebræ.

In the long cervical vertebræ of the Camel and Llama, the upper and lower transverse processes are not developed in the same perpendicular plane on the sides of the vertebræ, but at some distance from each other; the lower transverse processes (a, fig. 1, Pl. VI.; a, fig. 1, 3, 4, Pl. VII.) being given off from the lower part of the anterior extremity of the body of the vertebra; the upper ones (b, fig. 1, Pl. VI.; a, fig. 1, 3, 4, Pl. VII.) from the base of the superior arch near the posterior part of the vertebra, or from the sides of the posterior part of the body of the vertebræ. The extremities of these transverse processes do not become united together, but they either pass into each other at their base, or continue throughout life separated by an oblique groove (as in fig. 1, Pl. VI.) This groove would not, however, afford sufficient defence for the important arteries supplying those parts of the brain which are most essential to life; and, accordingly the vertebral arteries here deviate from their usual course, in order that adequate protection may be afforded to them in their course along the neck. From the sixth to the second cervical vertebræ inclusive in the *Auchenia*, and from the fifth to the second inclusive in the *Cameli*,* the vertebral arteries enter the vertebral canal itself, along with the spinal chord, at the posterior aperture in each vertebra, run forwards on the outside of the dura mater of the chord between it and the vertebral arch, and when they have thus traversed about two-thirds of the spinal canal, they perforate respectively the superior vertebral laminæ, and emerge directly beneath the anterior oblique or articulating processes, whence they are continued along with the spinal chord into the vertebral canal of the succeeding vertebra, and perforate the sides of the anterior part of the superior arch in like manner; and so on through all the cervical vertebræ until they reach the atlas, in which their disposition, and consequently the structure of the arterial canals, resemble those in other Ruminants.

The two cervical vertebræ of the Macrauchenia present precisely the structure and disposition of the bony canals for the vertebral arteries which are peculiarly characteristic of the Camelidæ among existing Mammalia. In Plate VI. fig. 2, the groove and orifices of the canal for the vertebral artery are shown in a section exposing the spinal canal: in Plate VII. figures 1 and 3 exhibit the orifices at the commencement of the arterial canals, as seen in a posterior view of the vertebræ; in figs. 2 and 4, the terminations of the same canals are shown, in the anterior view of the same vertebræ; the smaller figures (3 and 4) are taken from the fourth cervical vertebra of a Llama. The vertebræ of the Macrauchenia also closely resemble the middle cervical vertebræ of the Vicugna and Llama in their elongated form; approaching the Auchenial division of the Camelidæ, and deviating from the true Camels in the relations of the length of the body of

---

* In the seventh cervical vertebra of the Camel, as in many other Mammalia, there is no perforation in any part for the vertebral arteries. In a Vicugna, I find the same structure; but in a Llama, the side of the body of the seventh cervical vertebra is perforated longitudinally on the right side. In the Camel, the vertebral arteries pierce the sixth cervical vertebra, immediately below the superior transverse processes, and pass obliquely to the anterior aperture of the cervical canal, where they emerge beneath the anterior oblique processes, and then enter the spinal canal of the fifth cervical vertebra, as described in the text.

the vertebra to its breadth and depth, and in the much smaller size of the inferior processes. Excepting the Giraffe, there is no existing mammal which possesses cervical vertebræ so long as the Macrauchenia; but the cervical vertebræ of the Giraffe, differ in the situation of the perforations for the vertebral arteries, and in the form of the terminal articular surfaces, as will be presently noticed.

Both of the cervical vertebræ of the *Macrauchenia* here described, are of the same size, each measures six inches and a half in extreme length, two inches, ten lines in breadth, and two inches, four lines in depth. In the Giraffe and the Camelidæ, the spinous processes are thin laminæ of considerable extent in the axis of the vertebra, but rising to a very short distance above the level of the vertebral arch: the spinous processes have the same form in the corresponding vertebræ of the Macrauchenia, but present a still greater longitudinal extent; they commence at the interspace of the anterior oblique processes, and extend to opposite the base of the posterior oblique processes; the upper margin describing a gentle curve, as shown in fig. 1, Pl. VI. The transverse processes also present the form of slightly produced, but longitudinally extended, laminæ: their disposition is essentially the same as in the Camelidæ, but more nearly corresponds with the modifications presented by the Auchenix. The inferior transverse processes,—those which are alone developed in fish, but which are not present in any other vertebræ save the cervical, in mammalia,—these processes in the Macrauchenia are continued from the sides of the under surface of the anterior part of the body of the vertebra; their extremities being broken off, it cannot be determined how far they extended from the body of the vertebræ, but they gradually subside as they pass backwards: the superior transverse processes are continued outwards from the sides of the posterior part of the body of the vertebra, and gradually subside as they advance forwards along three-fourths of the body of the vertebra: they are not continued into the anterior and inferior transverse processes, as in the Vicugna, but are separated therefrom by a narrow and shallow groove. The articular, or oblique processes, closely resemble those of the Auchenia in form, and in the direction of the articular surfaces; those of the anterior processes looking inwards and a little upwards; those of the posterior, outwards and a little downwards.

In the Macrauchenia a small longitudinal process (*c*, fig. 2, Pl. VII.) is given off immediately below the base of the anterior oblique process; this structure is not observable in any of the cervical vertebræ of the Giraffe or Camelidæ.

In the form of the articulating surfaces of the bodies of the vertebræ the Macrauchenia deviates from the Giraffe and Camel, but resembles the Aucheniæ. In the Giraffe and Camel the anterior articulating surface is convex and almost hemispheric, the posterior surface is proportionally concave, so that the cervical vertebræ are articulated by ball and socket joints; yet not, as in most Reptiles, with intervening synovial cavities, but by the concentric ligamentous intervertebral substance characteristic of the Mammiferous class. In the Llama and Vicugna, the degree of convexity and concavity in the articular surface of the bodies of the cervical vertebræ is much less than in the Camels; and in consequence they carry their necks more stiffly and more in a straight line. In Macrauchenia the anterior articulating surface (fig. 2, Pl. VII.) presents a still slighter convexity than in the Llama (fig. 4, Pl. VII.), and the posterior surface (fig. 1, Pl. VII.) presents a correspondingly shallower concavity. The form of the extremities of the body of the vertebræ, especially of the posterior, is sub-hexagonal, the breadth being to the depth as eight to five. The sides and under part of the vertebræ are slightly concave; on the inferior surface there are two ridges, continued forwards from the posterior margin of the vertebra, each situated about an inch distant from the middle line; they converge as they pass forwards, and are gradually lost in the level of the vertebra; their greatest elevation does not exceed half an inch. In the Aucheniæ there is a longitudinal protuberance in the mesial line, instead of the two ridges. The two long cervical vertebræ of the Macrauchenia are also characterized by the maintenance of an almost uniform diameter of the body, both in its vertical and transverse extent; the cervical vertebræ of the Vicugna come nearest to them in this respect; those of the Camel deviate further in the large excavation at the under part of the body.

The long vertebral or spinal canal offers a slight enlargement at the two extremities; this structure which is generally in the ratio of the extent of motion of the vertebræ on each other is more marked in the Camel, where the form and mode of articulation of the bodies of the vertebræ are designed to admit of a free and extensive inflection of the cervical vertebræ; and the result of this structure is very obvious in the sigmoid flexure of the neck in the living animal. In the Auchenia, on the contrary, the neck is carried less gracefully erect and in an almost straight line, and the form of the vertebræ and the nature of their joints correspond, as we have seen, to this condition. From the length of the bodies of the cervical vertebræ of the Macrauchenia, and the almost flattened form of their anterior and posterior articular surfaces, I infer that the long neck in this singular quadruped must have been carried in the same stiff and upright position as in the Vicugna and Guanaco.

The following individual differences are observable in the two cervical vertebræ of the Macrauchenia;—in the posterior one the superior arch is wider and with thicker parietes, the body is more concave below, and the inferior transverse processes have a more lengthened origin.

Not a fragment of dorsal vertebræ, ribs or sternum, is included in the collection of the bones of the Macrauchenia; but fortunately seven lumbar vertebræ, forming a consecutive series of the same individual as that to which the cervical vertebræ belonged, were obtained, all more or less fractured, but all sufficiently perfect to demonstrate their true nature. These vertebræ, although not possessing such distinctive characters as the cervical, contribute by no means an unimportant element towards the illustration of the osteology of the Macrauchenia, and support the view which I have taken of its affinities; for, although, as will be seen from the structure of its extremities, this animal must be referred to the Order Pachydermata, yet no existing species of that order has more than six lumbar vertebræ; whilst among the Ruminants it is only in the Camel, Dromedary, Llama and Vicugna, that the lumbar vertebræ reach the number seven,— the same number which characterizes the extinct annectant species in question.

The dimensions of the vertebræ in the Macrauchenia present the same relations to the two cervical vertebræ above described, which the lumbar vertebræ of the Vicugna bear to the third, fourth, or fifth of its cervical vertebræ. But here we begin to discover modifications of form, in which the Macrauchenia deviates from the Camelidæ, and approaches the Pachyderms, as the Horse and Hippopotamus; and these indications become stronger as the vertebræ approach the sacrum.

In the Camel, as well as in the Horse and Hippopotamus, the bodies of the lumbar vertebræ diminish in vertical extent, or become flatter, as they approach the sacrum; but this character is more strongly marked in the Macrauchenia than in either of the above species. But in the Camelidæ the transverse processes of the lumbar vertebræ, are elongated, flattened, and narrow, resembling ribs, except that they are nearly straight; and this is more particularly the case with the transverse processes of the last lumbar vertebræ, which are the narrowest of all in proportion to their length, and stand freely out without touching the sacrum. The transverse processes of the lumbar vertebræ of the Giraffe resemble those of the Camel, but are relatively smaller and shorter. In the Hippopotamus the transverse processes of the lumbar vertebræ are much broader in proportion to their length than in any of the Ruminants, and they increase in breadth to the last lumbar vertebra, which presents in addition, the following characters; each transverse process sends off from its posterior margin a thickened and transversely elongated protuberance, which supports a flattened articular surface adapted to a corresponding surface on the anterior part of the transverse process of the first sacral vertebra: it likewise presents on its anterior edge a flattened and rough surface, which is closely attached by ligamentous substance to the opposite part of the transverse process of the penultimate lumbar vertebra. A similar structure exists in the last two lumbar vertebræ of the Rhinoceros, Tapir, and Horse. In the latter animal, anchylosis of these articulating surfaces of the lumbar and sacral vertebræ generally takes place with age, and, judging from the character of the same surfaces in the Hippopotamus, the motion of its lumbar vertebræ upon the sacrum may in like manner become ultimately arrested.

Now in the Macrauchenia, as in the Pachyderms above cited, the transverse processes of the last lumbar vertebræ are of considerable thickness and extent, and are joined by enarthrosis to the transverse processes of the sacrum; but the bony structure of these joints would indicate that they were not subject to be obliterated by anchylosis. The articular surfaces which project from the posterior part of the transverse processes of the last lumbar vertebræ present a regular and smooth concavity, adapted to a corresponding convexity in the transverse processes of the first sacral vertebra. These articulating surfaces have evidently been covered with smooth cartilage; they present a pretty regular transverse ellipsoid form. A view of the three joints by which, independently of the two oblique processes, the last lumbar vertebra of the Macrauchenia was articulated with the sacrum, is given in Plate VIII. fig. 1. The transverse processes of the posterior lumbar vertebra, besides their agreement with those of the Horse and Hippopotamus in the structure just described, also correspond with them in general form, and deviate remarkably from those of the *Camelidæ* in their great breadth.

It will be seen that the articulations on the body and transverse processes of the last lumbar vertebra of the Macrauchenia differ from the corresponding articular surfaces of the Horse, inasmuch as the middle surface is convex, while the two lateral ones are concave, and these are moreover relatively larger than either in the Horse or Hippopotamus: by this structure the trunk was more firmly locked to that segment of the vertebral column, which receives and transmits to the rest of the body the motive impetus derived from the hinder extremities, which are in all quadrupeds the chief powers in progression; while at the same time the shock must have been diminished by the great extent of interposed elastic cartilages; and a certain yielding or sliding motion would be allowed between the lumbar vertebræ and sacrum.

The anterior oblique processes of the lumbar vertebræ of the Macrauchenia

(fig. 4, Pl. VIII.) have concave articular facets turned towards, and nearly continued into, each other at their lower extremities; so as to form together a deep semilunar notch, into which the corresponding convex articular surfaces of the posterior oblique processes of the adjoining vertebra (fig. 3, Pl. VIII.) are firmly locked. In the close approximation of the two anterior concave articular facets, which are separated from each other only by a vertical ridge, and a rough surface of about three or four lines in breadth, the lumbar vertebræ of the Macrauchene resemble those of the Horse, and differ from those of the Camel-tribe and Ruminants generally, in which those surfaces are wider apart. In the hook-like form, however, of these articular processes the lumbar vertebræ of the Macrauchene differ from those of the Horse; and resemble those of many Ruminant species, and of the Anoplothere;[*] but the degree of concavity of the articulating surface is not so great in the Macrauchene. It would be interesting to determine the relations which the lumbar vertebræ of the Macrauchene bear to those of the Palæothere; but the indication which Cuvier gives of the single lumbar vertebra, of which he had cognizance in the latter genus[†] is too slight to enable me to

---

[*] Cuvier, Ossemens Fossiles, iii. p. 238.
[†] Loc. cit. p. 234.

enter upon the comparison.

The whole length of the lumbar region in the Macrauchene is twenty inches. When the bodies of these vertebræ are naturally adapted together, they form a slight curve, indicating that the loins of the Macrauchene were arched, or bent downwards towards the sacrum. That the lumbar vertebræ were rigidly connected together, or but slightly flexible, is evident from the flatness of the articular surfaces of the vertebral body, and by the circumstance of ossification having extended along the anterior vertebral ligaments, and produced an anchylosis between the fourth and fifth lumbar vertebræ; (fig. 2, c, Pl. VIII.) This kind of ossification is frequent in aged horses, and I have seen an example of a similar anchylosis of the lumbar vertebræ, by abnormal deposition of bone in their anterior ligaments, in the skeleton of a Hippopotamus preserved in the Senkenbergian Museum, at Frankfort.

In preparing the preceding account of the cervical and lumbar regions of the vertebral column of the Macrauchene, I have felt frequently a strong desire to enter into a comparison between them and the corresponding vertebræ of the extinct Pachyderms of the Paris Basin. Some of these, as the *Anoplotherium gracile*, in the length and slenderness of the cervical vertebræ, resemble both *Auchenia* and *Macrauchenia;* others, as the *Palæotherium minus*, and probably the rest of the genus, resemble the *Camelidæ* and *Macrauchenia* in having seven lumbar vertebræ. Cuvier points out the resemblance which the atlas of the Anoplothere bears to that of the Camel, and especially of the Llama;* but he expressly notices the existence of the canals for the vertebral artery in the fifth or sixth cervical vertebra of the *Anoplotherium commune*.† Do the cervical vertebræ—say from the third to the sixth inclusive—of the *Palæotherium* present an imperforate condition of their transverse processes, or exterior

---

* Loc. cit. p. 235

† Loc. cit. p. 237.

part of their sides? Cuvier, who seems not to have been aware of this peculiarity in the *Camelidæ*, merely notices the absence of these arterial foramina in the last cervical vertebra of the *Palæotherium minus*,‡ which, unfortunately for the comparison I am desirous of establishing, is that which most commonly presents this imperforate condition in the Mammalia generally. As, however, the cervical vertebræ of the Palæothere had the anterior articular surface of the body convex, and the transverse processes produced into descending laminæ, it is most probable that they corresponded with the cervical vertebræ of the typical Pachyderms in the condition of their arterial foramina.

The sacrum and ossa innominata in the present specimen of *Macrauchenia* are very imperfect; but sufficient is preserved to show that the sacrum was anchylosed to the ilia: the lower boundary of this anchylosis is marked below by an external ridge, and by vascular canals and grooves in the substance of the bone, as in the Hippopotamus. The body of the sacrum is lost, but the smooth articular convexities upon the transverse processes adapted to the articular depressions of the last lumbar vertebra are fortunately preserved.

The remains of the anterior extremity of our Macrauchenia include fragments of a left scapula; the proximal extremities of the anchylosed bones of the right antibrachium; the metacarpal and most of the phalangeal bones of the right fore-foot. The first-mentioned fragments, include the head and neck of the scapula, a small part of its body with the beginning of the spine, the coracoid process, and the nearly entire glenoid cavity. This articular surface (fig. 2, Pl. IX.) resembles in its general form, and degree of concavity, that of the Camel and Rhinoceros, and is deeper than in the Hippopotamus. The coracoid process is represented by a slightly produced rough, thick, and obtuse tuberosity, situated closer to the glenoid cavity than in the *Camelidæ* or *Rhinoceros*, and having almost the same relative position and

---

‡ Loc. cit. p. 235.

size, as in the *Palæotherium crassum*. The superior border or costa of the scapula presents much variety in the Ungulate quadrupeds with which we have to compare the Macrauchenia. In the Ruminants its contour forms behind the coracoid a concave sweep, which advances close to the spine of the scapula. In the Camel and Horse the marginal concavity is shallower, and the distance of the superior costa from the spine of the scapula is greater; the extent of the supra-spinal fossa increases in the true Pachyderms, and the Macrauchene agrees with them in this structure. In the Tapir, however, the contour of the superior costa is broken by a deep round notch immediately behind the coracoid: in the Hippopotamus this process arches in a slight degree backward over a corresponding but wider and shallower notch. In the *Palæotherium crassum* the concavity of the superior costa, behind the coracoid, is as slight as in the Rhinoceros; but in the Macrauchenia the superior costa of the scapula begins to rise or stretch away from the parallel of the spine, immediately behind the coracoid process. The modifications of the spine of the scapula which characterize respectively the Ruminants and Pachyderms have been clearly and concisely set forth by Cuvier, who at the same time points out the exceptional condition which the *Camelidæ* present in the production of the acromial angle. It was with peculiar interest and care, therefore, that I reunited all the fragments of the scapula of the Macrauchene, in the hope of gaining from this part of the skeleton as decisive evidence of an affinity to the Camel as the cervical vertebræ had afforded. It unfortunately happens, however, that the part of the scapula most important in this comparison is broken off; yet from this very circumstance, combined with a slight inclination forwards of the anterior margin of the spine immediately beneath the fractured acromion, and from the thickness of the fractured surface, we may infer that the acromial angle of the spine was more produced than in the ordinary Ruminants, although evidently in a less degree than in the Camel tribe. The Macrauchenia, however, surpasses these aberrant Ruminants, and equals the Pachyderms in the elevation and extent of its scapular spine: but this process commences about half an inch behind the glenoid cavity, and rises at once to the height of three inches above the plane of the scapula; in which structure we may trace the same tendency to the Ruminant type, as is manifested in the scapula of the Hippopotamus and Anoplotherium; for in most other Pachyderms the spine increases gradually from its extremities to the middle part. The anterior margin of the spine beneath the short acromion is perforated by an elliptical fissure measuring ten lines, by three lines. The extent of the spine which is preserved, measures eight inches and a half; it is a thin and nearly straight plate of bone, expanding into a thick and rugged upper margin, which slightly over-arches the inferior fossa, (fig. 1, Pl. IX.) In its general form and proportions the spine of the scapula in Macrauchenia presents the nearest resemblance to that of the Hippopotamus; but its origin is closer to the articular surface of the scapula than in this, or any other Pachydermal or Ruminant genus.

The portion of the antibrachium of the Macrauchenia which is preserved, presents a condition of the radius and ulna intermediate to those which respectively characterize the same bones in the Pachyderms and Camels. In the former the radius and ulna are separate bones, united in the prone position by ligament, yet so that the movement of supination cannot be performed; in the ordinary Ruminants they are partially joined by bony confluence, which rarely extends to the proximal extremities; in the Camel and Llama the anchylosis of the radius and ulna is so complete, that no trace of their original separation can be perceived, and the olecranon appears but as a mere process of the radius.

In the Macrauchenia the anchylosis of the radius and ulna is also complete, but the boundary line of the two originally distinct bones is very manifest, and the proportion which each contributes to the great articulating surface for the distal end of the humerus is readily distinguishable. About a sixth part of this surface is due to the head of the radius, which enters into the composition of the anterior and outer part of the articulation, and its extent is defined by a depressed line describing a pretty regular curve, with the concavity directed forwards and a little outwards, (*a*, fig. 1, Pl. X.) Just below the articular surface a strong triangular rugged protuberance projects from the front of the head of the radius, for the attachment of the tendon of the biceps. The line of separation of the radius and ulna is indicated on the inner side of the head of the radius by a deep and narrow fissure extending downwards from below the anterior part of the articulating surface; and on the outer side by a broad groove leading upwards to a deep pit near the proximal end of the antibrachium. We may see by the direction of the head of the radius which is thus defined, that it crosses obliquely in front of the ulna, as in the Elephant, Hippopotamus, and other Pachyderms, and that the bones are anchylosed in the prone condition: below this fissure and groove, which mark the interosseous line, the radius and ulna become blended together into one compact bone, which is flattened from before backwards, with a well marked ridge on the outer side; and excavated by a single medullary cavity, the compact walls of which present a general thickness of one-third of an inch.

The proximal articular surface or sigmoid cavity of the antibrachium, constituted as above described, resembles that of the Palæothere, Tapir, and the generality of the Pachyderms in having two depressions, instead of three, as in the Anoplothere, and Ruminants. The Hippopotamus has a slight tendency to the latter structure, which is also less marked in the Camel than in the ordinary Ruminants. In its general form the sigmoid cavity of the Macrauchene resembles that of the Hippopotamus more than that of the Camel. In the Camel this articular surface is traversed transversely by a broad, shallow, and slightly roughened tract, which divides the smooth surface of the joint into two parts, one forming the anterior horizontal surface due to the conjoined radius and ulna, the other forming the vertical concave surface on the anterior part of the base of the olecranon. In the Hippopotamus there is, as it were, an attempt at a similar division of the articulating surface at the proximal end of the antibrachial bones; a deeper and rougher depression encroaches upon the articulation from its outer side, but stops when it has reached half-way across. In the Macrauchenia the roughened surface, (*b*. fig. 1, Pl. X.) commencing also at the outside, extends only one-third of the way across the articular surface: it is, however, as shallow as in the Camel. The articular surface on the anterior part of the base of the olecranon is broader in the Hippopotamus than in the Camel; but in the Macrauchene it is twice as broad as in the Hippopotamus. The size of the olecranon in the Macrauchene exceeds that of the Hippopotamus, and *à fortiori* that of the Camel: indeed in its general magnitude the Macrauchenia must have fully equalled the largest Hippopotamus; but it no doubt had a more shapely, and less broad and bulky trunk. The olecranon of the Macrauchenia differs in shape, both from that of the Camel and Hippopotamus; it terminates above in a three-sided cone with an obtuse apex; and presents a well marked protuberance at the outer side of the base, which is not present in either the Camel or Hippopotamus. There is also a strong rugged ridge on the back part of the olecranon which makes an angle before sinking into the level of the ulna below.

The confirmation of the close affinity of the Macrauchenia to the Pachydermatous Order, which the structure of the cervical vertebræ alone might have rendered very doubtful, is afforded by the bones of the right fore-foot (Pl. XI.); these are fortunately in so perfect a condition, as to make it certain that this interesting quadruped had three toes on the fore-feet, and not more; and that the fully developed metacarpal bones are distinct, and correspond in number with the toes, and are not anchylosed into a single cannon bone, as in the Ruminants. The bones preserved are the metacarpals, proximal phalanges, and middle phalanges of each of the three toes, and the distal phalanx of the innermost toe.

The proximal end of the innermost metacarpal bone presents three articular surfaces; the middle facet is the largest, and the two lateral ones slope away from it at an angle of 45°. The middle facet is broad and slightly convex in front, narrow and concave behind; the distal articular surface of the trapezoides must have corresponded with this surface; the outer facet is narrow, flat, extends from the fore to the back part of the head of the bone, and must have been adapted to a corresponding surface on the os magnum; the inner facet is the smallest, presents a triangular form, and is situated towards the back part of the head of the metacarpal bone; it indicates the existence of a rudimental metacarpal bone, or vestige of a pollex. Below the outermost of the lateral surfaces there is a crescentic articular surface with its concavity directed outwards and downwards (fig. 2, Pl. XV.), against which a corresponding convex articular surface of the middle metacarpal abuts, (fig. 3, Pl. XV.) External to this surface the proximal end of the middle metacarpal bone presents two articular surfaces for the carpus; the larger one, which was adapted to the os magnum, is horizontal, broad and convex before, narrow and concave behind; the outermost facet is a small triangular surface inclined downwards to the level of the articulating surface of the outermost metacarpal. It also presents a posterior vertical articular surface for a sesamoid bone. The proximal extremity of the outer metacarpal bone is joined to the middle metacarpal, not by one semilunar surface, but by two separate articulations of small size (fig. 4 and 5, Pl. XV.); it presents a single large slightly convex articular surface for the os magnum, of an irregular semicircular form, with the convexity of the curve turned outwards.

The metacarpus increases in breadth as it approaches the phalanges; the two lateral metacarpals bending slightly away from the middle one, and expanding towards their distal extremities: the middle bone presents a symmetrical figure except at its proximal extremity (fig. 2, Pl. XI.) The distal articulating facet of each of the metacarpal bones extends so far upon both the anterior and posterior surfaces as to describe more than a semi-circle (fig. 3, Pl. XI.); in the two lateral metacarpals it is traversed throughout by a longitudinal convex ridge dividing it into two equal lateral parts; the ridge is most produced on the posterior half of the joint (fig. 4, Pl. XI.): in the middle metacarpal this ridge subsides before it reaches the anterior part of the articular surface.

The proximal extremity of the middle proximal phalanx presents a posterior notch corresponding to the above partially developed ridge: the proximal extremities of the lateral phalanges are traversed by a middle longitudinal depression, and two lateral shallow concavities (fig. 6, Pl. XI.;) but these are of such an extent as to be in contact with only a part of the convexity above, which therefore was doubtless adapted to a sesamoid bone on each side of the longitudinal ridge. The structure of the above described joints proves that the motion of the toe upon the metacarpus was much freer and more extensive than in the Rhinoceros, which is the only existing Ungulate mammal which presents the tridactyle structure in the fore-foot. In this species the metacarpo-phalangeal articulations exhibit only a slight trace of the longitudinal ridges and grooves which are confined to the posterior part of the joint; these are more developed in the *Camelidæ;* but the Hog and Horse in this respect approach nearer to the Macrauchene, though the structure of the metacarpo-phalangeal joints in the Hog falls far short of the compactness and strength combined with freedom of play in flexion and extension which distinguish those of the Macrauchene. The *Palæotherium medium* most resembles the Macrauchene in the structure of the trochlear metacarpo-phalangeal joints; but both in this species,* and the *Pal. crassum*† the articular surface at the distal end of the metacarpal is relatively narrower than in the Macrauchenia;

---

* See Ossem. Fossiles, Pl. XX. fig. 3.
† Loc. cit. Pl. XXII. fig. 6.

moreover all the species of the extinct Palæothere differ from the Macrauchene in the greater size and strength of the middle as compared with the lateral metacarpals.

The articulation at the distal extremity of the proximal phalanges (fig. 5, Pl. XI.) is simple, and not divided into two pulleys by a longitudinal ridge; it is slightly concave from side to side; but in its extent upon the anterior and posterior surfaces of the bone indicates a freedom of flexion and extension of the toes, which harmonizes with the structure of the joint above.

The proximal articulating surfaces of the second phalanges (fig. 7, Pl. XI.) corresponds of course to those to which they are adapted; they are, however, characterized by sending upwards an obtuse process from the middle of their anterior margin. The distal articulating surfaces (fig. 8, Pl. XI.) resemble those of the proximal phalanges, but extend further upon the back part of the phalanx than the front, indicating the more horizontal position of the second phalanges.

The last phalanx, does not resemble the neatly defined ungulate phalanges of the Ruminantia, and Solipedia, but has the irregular form characteristic of those of the Pachydermata. It is wedge-shaped, broader than it is long, with a rugged surface, except where it plays upon the distal end of the second phalanx, where it is slightly concave in one direction, and convex in the other, (figs. 7 and 9, Pl. XI.) A portion of this phalanx extends backwards behind the articular surface, as in the corresponding bone of the Palæothere and Rhinoceros.

The femur of the Macrauchenia (fig. 1, Pl. XII.) is full two feet in length, and consequently longer than in any known Camel or Rhinoceros; as compared with its transverse diameter it is much longer than the femur of the latter animal: in the proportion of its breadth to its length, and the expansion of its extremities as compared with the diameter of the shaft, it more resembles that of the Camel. The femur of the Giraffe deviates from that of the Macrauchenia in the excessive expansion of its distal extremity. But the most striking evidence deducible from this bone, of the affinity of the Macrauchenia to the true Pachydermatous type is afforded by the evident traces of a third trochanter, the outline of which is conjecturally restored in the figure. Of the Pachyderms which have this characteristic structure, the extinct Palæothere offers the nearest resemblance to the Macrauchene in the general form and structure of the femur.

The head of the femur in the Macrauchene (fig. 2, Pl. XII.) presents the form of a pretty regular hemisphere; it is less flattened above, and is directed more obliquely inwards than in the Palæothere: the neck supporting it does not project so far from the shaft as in the Palæothere or Tapir, but farther than in the Camel. The great trochanter rises above the level of the head; in which structure and in the depression between the head and trochanter, the femur of the Macrauchene offers a character intermediate between the Tapir or Palæothere, and the Camel. The lesser trochanter is a slight projection from a ridge of bone which is continued from the under part of the head of the femur to the inner surface of the shaft. In the Palæothere the lesser trochanter is situated more towards the posterior surface of the femur; so that, in this particular, the Macrauchene approaches nearer to the Camel. Cuvier makes no mention of the condition of the depression for the *ligamentum teres* in the Palæothere. Among existing ordinary Pachyderms the Hippopotamus presents no trace of the insertion of a *ligamentum teres* in the head of the femur; in the Camel the place of its insertion is indicated by a well-marked circumscribed pit; in the Tapir a similar circular depression is situated close to the inferior margin of the articular convexity. The ligament was undoubtedly present in Macrauchenia, but the place of its insertion is a broad and deep notch leading from the under and back part of the head of the bone a little way into its articular surface: this I regard as another of those interesting transitional structures with which the remains of the Macrauchenia, few and imperfect though they unfortunately are, so freely abound.

The femur of Macrauchenia, in the flatness of the back part

of its neck, and the elongated form of the post-trochanterian depression, resembles that of the Camel rather than that of the Palæothere; and the same resemblance is shown in the cylindrical figure, straightness, and length of the shaft. The depth of the trochanterian depression, and the incurvation of the strong ridge continued downwards from the great trochanter are individual peculiarities in the Macrauchenia.

A great part of the third trochanter is broken off; but from the remains of its base we see that it had the same relative size as in the Palæothere; but it is situated at the middle of the shaft of the femur, and consequently lower down than in the Palæotheres and Tapirs. In the general form and relative size of the condyles at the distal extremity of the femur (fig. 3, Pl. IX. and XII.) the Macrauchene is intermediate to the Camel and Palæothere, but resembles more the latter. In the articular surface for the patella, it deviates somewhat from the Palæothere, having this part longer in proportion to its breadth, more regularly and deeply concave from side to side, and with its lateral boundaries more sharply defined. In all these points the Macrauchene approaches the Camel: the same affinity is shown in the protuberance above the inner condyle; but in the extent of the posterior projection of this condyle (fig. 3, Pl. IX.) it exceeds the Camel and Palæothere, and displays an intermediate structure between these species and the Hippopotamus.

There is a rough crescentic depression above the outer condyle where the linea aspera begins to diverge; the corresponding depression is deeper in the Hippopotamus, while in the Camel it is represented by a roughened surface only, which is not depressed. In the fossa between the rotular articulation and the external condyle the Macrauchene resembles the Camel: the interspace of thecondyles is relatively wider than in the Camel, and the process above the inner condyle is more angular; in both these respects the Macrauchene inclines towards the Palæothere.

In the structure of the bones of the leg of the Macrauchenia we find the same transitional character which is afforded by the definable limits of the anchylosed bones of the fore-arm. In the Pachyderma the fibula is an entire and distinct bone. In the Ruminantia, with the exception of the small Musk-deer, and, in an inferior degree, the Elk, the fibula appears only as a short continuous process sent down from the under part of the external condyle of the tibia. In the Camel tribe the only trace of the fibula in the bones of the leg, is this process in a still more rudimental state. In the Macrauchenia the fibula is entire, but is confluent with the tibia through nearly its whole extent: the proximal part of the fibula is well defined; its head is anchylosed to the outer condyle of the tibia, but the shaft is continued free for the extent of nearly two inches, and then again becomes confluent with the tibia, forming apparently the outer ridge of that bone. About five inches from the distal end of the tibia this outer ridge becomes flattened by being, as it were, pressed against the tibia, and the anterior and posterior edges are raised above the level of the tibia; beyond this part the limits of the fibula begin again to be defined by deep vascular grooves. The outer side of the distal end of the fibula is excavated by a broad tendinous groove. The fibula and tibia are distinct bones in both the Palæothere and Anoplothere, as in the Pachyderms. It is to the former genus, however, especially *Pal. magnum*, that the Macrauchene presents the nearest approach in the general form of the tibia, the principal bone of its leg: but in the Macrauchene the tibia is relatively shorter, and thicker, and is straighter and less expanded at its extremities, especially the upper one, than in any of the Palæotheres.

The mesial boundaries of the two superior articulating surfaces of the tibia are raised in the form of ridges, which are separated by a deep groove; of these ridges the external is the highest, as in *Pal. magnum*: but the articular surfaces in the Macrauchene slope away from these ridges more than in the Palæotheres. The rotular or anterior tuberosity of the tibia is more produced, and rises higher than in the Palæotheres; the ridge continued downwards from this process is more marked in the Macrauchene, and its limits are better defined: the shaft of the tibia below the ridge is also more flattened in the antero-posterior direction than in the Palæothere. The configuration of the back part of both proximal and distal extremities of the tibia are so clearly and accurately given in figures 2 and 3, Pl. XIII., as to render verbal description unnecessary. Neither the text nor the figures in the 'Ossemens Fossiles' afford the means of pursuing the comparison between the Macrauchene and Palæothere in these particulars; and I proceed, therefore, to the consideration of the inferior articulating surface of the bones of the leg (fig. 4, Pl. XIII.)

Since, of the hind foot, we possess in the present collection only a single tarsal and metatarsal bone, the structure of the distal articular surface of the tibia is attended with peculiar interest, because we are taught by Cuvier that it reveals to us in the Ungulate animals the didactyle or tridactyle structure of the foot. In the Ruminants this articular surface is nearly square, and extended transversely between two perpendicular malleoli, while in the Pachyderms with three toes to the hind-foot the articular surface of the tibia is oblique, and is divided into two facets between the perpendicular malleolar boundaries. Now in the Macrauchenia, although the two bones of the leg are anchylosed together, the extent of that part of the tarsal articular surface which is due to the tibia is indicated, as in the case of the radius in the joint of the fore-arm, by a groove; and we are thus able to compare this surface with the distal articular surface of the tibia in the Palæothere and Anoplothere. It presents in the Macrauchenia a very close resemblance with that of the *Palæotherium magnum*,[*] being divided into two facets by a convex rising, which traverses the joint from behind forwards; but the ridge is narrower, the internal facet somewhat deeper, and the external oblique surface rather flatter than in the three-toed Palæothere. In the portion of the tarsal articular surface

---

[*] See Ossem. Foss. iii. Pl. XXVI. fig. 5.

due to the fibula, we find, however, a more marked deviation from the Palæothere, and an interesting correspondence with the Anoplothere, in the inferior truncation and horizontal articular surface which is continued upon the lower extremity of the fibula, at right angles with the vertical malleolar facet which forms the outer boundary of the trochlea of the astragalus: this articular surface unerringly indicates a corresponding articular projection in the calcaneum, which, therefore, although the bone itself does not form part of the present collection, we may conclude to differ from the calcaneum of the Palæotherium, and to resemble that of the Anoplotherium, in this particular at least.

The valuable indication which the distal articular surfaces of the anchylosed tibia and fibula have given of the correspondence of the hind-foot with the fore-foot of the Macrauchenia, in regard to the number of the toes, receives ample confirmation from the astragalus, which, of all the bones in the foot, is the one that an anatomist would have chosen, had his choice been so limited, and which most fortunately has been secured by Mr. Darwin, in a very perfect state, in the present instance. I have compared this astragalus with that of the Giraffe, and other Ruminants, the Camel, the Anoplothere, the Horse, the Hog, the Hippopotamus, Rhinoceros, Tapir, and Palæothere: it is with the Pachyderms having three toes to the hind-foot, that the Macrauchenia agrees in the main distinguishing characters of this bone; its anterior articular surface, for example, is simple, and not divided into a double trochlea by a vertical ridge: lastly, it is with the astragalus of the Tapir and Palæothere that it presents the closest correspondence in the general form and the minor details of structure, and with these Pachyderms, therefore, I shall chiefly limit the comparison of the Macrauchenia, in regard to the bone in question. If the upper or tibial articular surface (fig. 5, Pl. XIV.) be compared with that in the *Palæotherium magnum* (Ossem. Foss. Pl. LIV. fig. 2,) it will be seen, that the general direction of that surface is more parallel with the axis of the bone in Macrauchenia. In the Palæotherium it is turned a little towards the outer or fibular side, and in the Tapir the general direction of the same surface is placed still more obliquely. The anterior border of this articulating surface is broken by a semicircular notch in the Palæothere; in the Tapir it describes a gentle concave curve, and the Macrauchene resembles the Tapir in this respect. The chief difference between the astragalus of the Tapir and the Palæothere, when viewed from above, obtains in the relative length of the bone, anterior to the tibial articulating surface: the Macrauchene presents, in this respect, an intermediate structure, but differs from both in the greater extent of the tibial side of this part of the astragalus.

If we next direct attention to the anterior or scaphoid articular surface, (fig. 3, Pl. XIV.) and compare it with that of the *Palæotherium magnum,* (fig. 4, Pl. liv, Ossem. Foss.) it will be seen, that it presents in the Macrauchenia an oval, and in the Palæotherium an irregular quadrangular form: in the Macrauchenia, this surface is uniform or undivided, and is gently convex, except at its lower part; while in the Palæothere it is divided by an oblique ridge into a broad internal facet for the scaphoid bone, and a narrow internal surface for articulation with the os cuboides; the larger surface is also concave transversely, and slightly convex vertically: in the Tapir, the anterior surface of the astragalus deviates still further from that of the Macrauchenia, both in general form, and in the proportion of the cuboidal facet. In the didactyle Anoplotherium, Camel, and true Ruminants, where the cuboides presents a large relative size, a still greater proportion of the anterior surface of the astragalus is devoted to the articulation with this bone, and is separated from the scaphoid surface by a well-developed vertical ridge. The Macrauchenia presents, therefore, the extreme variation from this type;—and should the entire tarsus hereafter be discovered, it will doubtless be found, that the os cuboides is articulated posteriorly to the os calcis exclusively.

The external surface of the astragalus of the Macrauchene, (fig. 1. Pl. XIV,) is longer in proportion to its vertical extent than in the Tapir or Palæothere: the articular surface for the fibular malleolus is less curved. Between this surface and the anterior facet the bone is excavated by a deep notch, both in the Tapir and Palæothere; but in the Macrauchenia by a gentle concavity. Beneath the malleolar articular smooth surface in the Palæothere there is a deep pit; in the Tapir a shallow one; but in the Macrauchenia we observe only a smooth and slightly convex triangular surface. If we compare the inner surface of the astragalus in these three animals, we shall find the existing Tapir again forming a transition between the two extinct genera. In the Palæothere, a round protuberance projects from the anterior part of this surface: in the Tapir, we observe a gentle rising of the bone in the same part, while in the Macrauchene (fig. 2) the surface of the bone is level at this part. The margin of the tibial malleolar articular surface, which is very slightly raised in the Macrauchene, is more developed in the Tapir, and still more so in the Palæothere, where it forms a ridge, overhanging the rough outer side of the bone. Near the lower part of this surface we observe a small but deep depression in the Palæothere; there is a shallower one in the corresponding part in the Tapir; and the depression is still wider and shallower in the Macrauchenia. In the Palæothere the astragalus articulates by three surfaces with the os calcis, posteriorly by a large concave surface, externally by a longitudinal sub-elliptic surface, and anteriorly by a thin transverse facet: in the Macrauchene (fig.4) two only of these surfaces are present, viz. the concave and the longitudinal one, the anterior transverse surface being wanting: in the Tapir, the transverse surface is present, but is confluent with the longitudinal one. The posterior surface is relatively larger and deeper in the Macrauchene than in the Palæothere, and approaches nearer to the triangular than the oval form: the longitudinal surface is placed more obliquely, and is truncated anteriorly. In the Tapir this surface is confluent with the scaphoid articular surface, but it is separated therefrom by a narrow strip of bone in both the Palæothere and Macrauchene. It is satisfactory to find in the bone, which marks most strongly the affinity of *Macrauchenia* to *Palæotherium*, so many easily recognizable differences, because the structure of the cervical vertebræ in the latter genus is too imperfectly known, to allow us to predicate confidently a distinction between it and *Macrauchenia* in that particular; the difference, however, which they present in the condition of the bones of the fore-arm and leg, forbids their being considered as generically related.

There remains to be noticed only a single fractured metatarsal bone (fig. 1. Pl. XV.) This, from its bent and unsymmetrical figure, is evidently not a middle one, and having the side of the proximal end, which was articulated to the adjoining metatarsal in a nearly perfect state, it enables us to refer it with certainty to the hind-foot, since it does not agree with any of the corresponding surfaces at the proximal extremities of the metacarpal bones. It remains then to be determined, whether it is an external metatarsal of the right-foot, or an internal one of the left-foot, the general curvature of these being in the same direction. With neither of these bones in the Tapir does our metatarsal agree, since it has but one articular facet on the lateral surface of its proximal end, while the outer metatarsal of the right-foot of the Tapir, with which, in other respects, it most closely corresponds, has two articular surfaces. In the cast of a hind-foot of a Palæothere, I find that the outer metatarsal bone closely agrees with this metatarsal bone of the Macrauchene, in the structure just alluded to: the articulation with the middle metatarsal being by a single sub-oval facet, which stands out a little way from the surface of the bone: the articular surface in the Macrauchene presents a similar form and condition, and is similarly situated to that in the Palæothere, being at the posterior part of the lateral surface, and a little below the superior or tarsal articular surface. The bone expands towards its distal end, which corresponds in structure with those of the two lateral metatarsals in the fore-foot, in being completely divided into two trochlear surfaces by a well-developed median ridge, and in having the posterior half of this ridge suddenly produced, so as to project about two lines further from the trochlear surface than the anterior part of the same ridge. In both the Tapir and Palæothere this anterior part of the ridge is wholly suppressed, and the posterior is much more feebly developed than in the Macrauchenia. The metatarsal bone here described is of exactly the same length with the internal metacarpal bone, and proves, in conjunction with the proportions of the astralagus, that the fore and hind feet of the Macrauchenia were of equal size.

Thus then we obtain evidence, from a few mutilated bones of the trunk and extremities of a single representative of its race, that there once existed in South America a Pachydermatous quadruped, not proboscidian, which equalled in stature the Rhinoceroses and Hippopotamuses of the old world. But this, though an interesting and hitherto unsuspected fact, is far from being the sum of the information which is yielded by these fossils. We have seen that the single ungueal phalanx bespeaks a quadruped of the great series of *Ungulata*, and this indication is corroborated by the condition of the radius and ulna, which are fixed immoveably in the prone position. Now in the Ungulated series there are but two known genera,—the Rhinoceros and Palæotherium,—which, like the quadruped in question, have only three toes on the fore-foot. Again, in referring the Macrauchenia to the Tridactyle family of Pachyderms, we find, towards the close of our analysis, and by a detailed comparison of individual bones, that the Macrauchenia has the closest affinity to the Palæotherium.

But the Palæotherium, like the Rhinoceros and Tapir, has the ulna distinct from the radius, and the fibula from the tibia; so that even if the Parisian Pachyderm had actually presented the same peculiarities of the cervical vertebræ as the Patagonian one, it would have been hazardous, to say the least, while ignorant of the dentition of the latter, to refer it to the genus *Palæotherium*.

Most interesting, indeed will be the knowledge, whenever the means of obtaining it may arrive, of the structure of the skull and teeth in the Macrauchenia.

Meanwhile, we cannot but recognise, in the anchylosed and confluent state of the bones of the fore-arm and leg, a marked tendency in it towards the Ruminant Order, and the singular modifications of the cervical vertebræ have enabled us to point out the precise family of that order, with which the Macrauchenia is more immediately allied.

In first demonstrating this relationship, it was shown in how many particulars the *Camelidæ*, without losing the essential characters of Ruminantia, manifested a tendency to the Pachydermatous type; and the evidence which the lost genera, *Macrauchenia* and *Anoplotherium*, bear to a reciprocal transition from the Pachyderms to the Ruminants, through the *Camelidæ*, cannot but be viewed with extreme interest by the Zoologist engaged in the study of the natural affinities of the Animal Kingdom.

The Macrauchenia is not less valuable to the Geologist, in reference to the geographical distribution of animal forms. It is well known how unlooked-for and unlikely was the announcement of the existence of an extinct quadruped entombed in the Paris Basin, whose closest affinities were to a genus, (*Tapirus*,) at that time, regarded as exclusively South American. Still greater surprise was excited when a species of the genus *Didelphys* was discovered to have co-existed in Europe with the *Palæotherium*.

Now, on the other hand, we find in South America, besides the Tapir, which is closely allied to the Palæothere, — and the Llama, to which the Anoplothere offers many traces of affinity, — the remains of an extinct Pachyderm, nearly akin to the European genus *Palæotherium:* and, lastly, this Macrauchenia is itself in a remarkable degree a transitional form, and manifests characters which connect it both with the Tapir and the Llama.

ADMEASUREMENTS OF THE BONES OF THE MACRAUCHENIA.

	Inches.	Lines.
Length of third (?) cervical vertebra	7	9
Vertical diameter of ditto	4	0
Do. do. of body of ditto	2	3
Transverse diameter of ditto	3	3
Vertical diameter of spinal canal	1	...
Length of fourth lumbar vertebra	5	5
Vertical diameter of body of ditto	2	9
Transverse diameter of ditto	2	10
Vertical diameter of spinal canal	1	1
Transverse ditto ditto*	1	6

* This diameter increases rapidly in the posterior lumbar vertebræ, in correspondence with the enlargement of the spinal chord, which gives off the great nerves of the hinder extremities.

	Inches.	Lines.
Transverse diameter of last lumbar vertebra	9	...
Ditto do. of body of ditto	2	2
Vertical diameter of ditto	1	3
Entire length of lumbar region of vertebral column	20	
Vertical diameter of glenoid cavity of scapula	3	...
Transverse ditto ditto ditto	2	10
Elevation of spine of scapula	3	5
Vertical diameter of proximal articular surface of fore-arm	3	6
Transverse ditto ditto ditto	3	5
Height of olecranon	5	3
Greatest diameter of its base	2	...
Circumference of proximal end of anchylosed radius and ulna	11	10

Entire length of inner toe of fore-foot, inclusive of metacarpal bone ..	13	...
Breadth of proximal end of metacarpus	3	8
Do. distal end of ditto	5	4
Length of inner metacarpal bone	7	6
Do. middle ditto	8	...
Do. outer ditto	7	...
Do. inner proximal phalanx	3	6
Do. middle ditto	2	10
Do. outer ditto	3	4
Do. inner middle phalanx	2	...
Do. middle ditto	2	3
Do. inner distal phalanx *	1	...
Do. the femur	24	...
Diameter of base of articular surface of the head of ditto	3	6
Greatest diameter of proximal end	7	...
Do. of distal end	6	3
Circumference of middle of shaft	8	...
Length of tibia	18	
Greatest diameter of proximal end	5	7
Do. of distal end, including fibula	4	4
Circumference of middle of shaft	9	...
Length of metatarsal bone†	7	4

* The relative breadth of these bones is shown in the figures of the fore-foot, Pl. XI.
† The figures in Pl. XIV. preclude the necessity of giving the admeasurements of the astragalus.

DESCRIPTION OF A FRAGMENT OF A CRANIUM OF AN EXTINCT MAMMAL, INDICATIVE OF A NEW GENUS OF EDENTATA, AND FOR WHICH IS PROPOSED THE NAME OF

GLOSSOTHERIUM.

"LA première chose à faire dans l'étude d'un animal fossile, est de reconnaitre la forme de ses dents molaires; on détermine par-là s'il est carnivore ou herbivore;" says Cuvier, at the commencement of that series of splendid chapters in which the restoration of the extinct Pachyderms of the Paris Basin is recorded. In the present case, however, as in that of the Mammiferous animal whose fossil remains we were last considering, the important organs, to which Cuvier directs our first attention, are wanting. Nor are there here, as in the Macrauchenia, any remains of the locomotive extremities to compensate for the deficiency of teeth, and guide us into the right track of investigation and comparison. The animal, the nature and affinities of which are the subject of the following pages, is, in fact, represented in Mr. Darwin's collection, by nothing more than a fragment of the cranium.

This fragment, which was found in the bed of the same river, (see p. 16,) in Banda Oriental, with the cranium of the Toxodon, includes the parietes of the left side of the cerebral cavity, the corresponding nervous and vascular foramina, the left occipital condyle, a portion of the left zygomatic process, and, fortunately also, the left articular surface for the lower jaw. The importance of this surface in the determination of the affinities of a fossil animal has been duly appreciated, since the relations of the motions of the lower jaw to the kind of life of each animal were pointed out by Cuvier; but yet we should be deceived were we to establish, in conformity with the generalization enunciated by

Cuvier,* our conclusion, from this surface, of the nature of the food of the extinct species under consideration; for the glenoid cavity is so shaped as to allow the lower jaw free motion in a horizontal plane, from right to left, and forwards or backwards, like the movements of a mill-stone; and, nevertheless, I venture to affirm it to be most probable, that the food of *Glossotherium* was derived from the animal and not from the vegetable kingdom; and to predict, that when the bones of the extremities shall be discovered, they will prove the Glossothere to be not an ungulate but an unguiculate quadruped, with a fore-foot endowed with the movements of pronation and supination, and armed with claws, adapted to make a breach in the strong walls of the habitations of those insect-societies, upon which there is good evidence in other parts of the present cranial fragment, that the animal, though as large as an ox, was adapted to prey.

We perceive, in the first place, looking upon the base of this portion of skull, a remarkable cavity, situated immediately behind the tympanic bone, of nearly a regular hemispherical form, an inch in diameter (fig. 2, *b*, Pl. XVI). The superficies of this cavity appears not to have been covered with articular cartilage, for it is irregularly pitted with many deep impressions; and I conclude, therefore, that it served to afford a ligamentous attachment to the styloid element of a large *os hyoides*. With this indication of the size of the skeleton of the tongue, is combined a more certain proof of the extent of its soft, and especially its

---

* "Comme le genre de vie de chaque animal est toujours en rapport avec les mouvements dont sa mâchoire est susceptible, on retrouve dans la conformation des surfaces destinées à l'articulation, les particularités qui semblent le déterminer d'avance. Ainsi dans les animaux qui vivent de chairs, substances filamenteuses qui ne peuvent être écrasées, mais seulement coupées et dechirées, le mouvement de la mâchoire inférieure ne peut s'exécuter que de haut én bas. Dans les herbivores, les frugivores et les granivores, comme le principal mouvement est celui de broiement pour écraser, comprimer les herbes et les fruits, pour briser les grains et les réduire, en pâte, le mouvement des mâchoires se fait encore de droite à gauche, et réciproquement, on en même temps, de devant en arrière, en un mot, dans un plan horizontal autant que dans un vertical: les uns représentent des ciseaux, les autres des meules de moulin."

muscular parts, in the magnitude of the foramen, for the passage of the lingual or motor nerve (*c.* fig. 2 and 3). This foramen, (the anterior condyloid,) in the present specimen, is the largest of those which perforate the walls of the cranium, with the exception of the foramen magnum; it is fully twice the size of that which gives passage to the second division of the fifth nerve; its area is oval, and eight lines in the long diameter, so that it readily admits the passage of the little finger.

It is only in the Ant-eaters and Pangolins that we find an approximation to these proportions of the foramen for the passage of the muscular nerve of the tongue; and the existing Myrmecophagous species even fall short of the larger fossil in this respect. Some idea of the size of the lingual nerve, and of the organ it was destined to put in motion, may be formed, when it is stated that the foramen giving passage to the corresponding nerve in the Giraffe,—the largest of the Ruminants, and having the longest and most muscular tongue in that order,—is scarcely more than one-fourth the size.

With these indications of the extraordinary development of the tongue, we are naturally led, in order to carry out a closer and more detailed comparison of the fossil in question, to that group of mammalia in which the tongue plays the chief part in the acquisition of the food. The size, form, and position of the occipital condyle,—the magnitude of the occipital foramen, (which must here have somewhat exceeded three inches in the transverse diameter,)—the slope of the occipital surface of the cranium from below, upwards and forwards, at an angle of 60° with the base of the cranial cavity—each and all attest the close affinities of the present animal to the Edentata. More decisive evidence of the same relationship will be adduced from the organization of other parts of the cranium. The glenoid articular surface (*a*, fig. 2, Pl. XVI.) is an almost flattened plane, wider in the transverse than in the longitudinal direction; and, as in the genera *Myrmecophaga* and *Manis*, it is not defended behind by any descending process. In its general form it resembles the glenoid cavity of *Orycteropus* more than that of the preceding Edentates; but, in *Orycteropus*, the articulation is defended posteriorly by a descending process of the zygoma, and it is also situated relatively closer to the os tympanicum.

Had the *Glossotherium* teeth ? The extent of the temporal muscle, which is indicated by the rugged surface of the temporal fossa, and by the well-marked boundary, formed by a slightly elevated bony ridge, which extends to near the line of the sagittal suture, together with the size of the zygomatic portion of the temporal bone, and the remains of the oblique suture by which it was articulated to the malar bone, enables me to answer this question confidently in the affirmative. They will probably be found to be molar teeth of a simple structure, as in the Orycteropus.

The evidence just alluded to of the existence of an os malæ is interesting, because this bone is wanting in the Pangolins; and its rudimental representative in the true Ant-eaters does not reach the zygomatic process of the temporal bone, which consequently has no articular or sutural surface at its anterior extremity. In the presence, therefore, of the surface for the junction of the os malæ, and the consequent evidence of the completion of the zygomatic arch, we learn that the Glossothere was more nearly allied to the Armadillos and Orycterope. That its affinity to the latter genus was closer than to the Armadillos we have most interesting evidence in the form and loose condition of the tympanic bone: it is represented of the natural size at fig. 4, Pl. XVI. Through the care and attention devoted to his specimens by their gifted discoverer, this bone was preserved *in situ*, as represented at *d*, fig. 1; but it had no osseous connection with the petrous or other elements of the temporal bone, and could be displaced and replaced with the same ease as in the *Orycterope*. This bony frame of the membrana tympani, in the Glossothere, describes rather more than a semicircle, having the horns directed upwards; it has a groove, one line in breadth, along its concave margin, for the attachment of the ear-drum, and sends down a rugged process, half an inch long, from its lower margin. In the *Dasypodes* and *Myrmecophagæ*, the tympanic bone soon becomes anchylosed with the other parts of the temporal; it is only in *Orycteropus*, among the existing insectivorous *Bruta* or *Edentata*, that it manifests throughout life the fœtal condition of a distinct bony hoop, deficient at the upper part. The os tympanicum of *Orycteropus*, however, differs from that of

Glossotherium, in forming part of the circumference of an ellipse, whose long axis is vertical; and in sending outwards, from its anterior part, a convex eminence, which terminates in a point directed downwards and forwards.

Such appear to be the most characteristic features of the cranial fragment under consideration, in which we have found, that the articular surface for the os hyoides throws more light upon the nature of the animal of which it is a part, than even the glenoid cavity itself. There now remains to be described as much of the individual characters of the constituent bones as the specimen exhibits.

The occipital bone, besides forming the posterior and part of the inferior parietes of the cranium, extends for about half an inch upon the sides, where the ex-occipital element is articulated by a vertical suture with the mastoid element of the temporal: this suture is situated in a deep and well-marked muscular depression (*e*, fig. 1), measuring three inches in the vertical, and upwards of one inch in the transverse direction. The other sutures, uniting the occipital to the adjoining bones, are obliterated. The breadth of the occipital region must have exceeded the height of the same by about one-third. The condyle extends nearly to the external boundary of the occipital aspect of the cranium; there is situated, external to it, only a small ovate, rounded and smooth protuberance. The slightly concave surface of the occipital plane of the cranium is bounded above by a thick obtuse ridge, the muscular impressions are well sculptured upon it. It is traversed transversely at its upper third by a slightly elevated bony crest; and the surface below this ridge is again divided by a narrower intermuscular crest, which runs nearly vertically, at about an inch and a half from the external boundary of the occipital plane. As a similar crest must have existed on the opposite side, the general character of the occipital surface in the Glossothere would resemble that of the Toxodon. A similar correspondence may be noticed in the terminal position of the condyle, and the slope of the occipital plane.

Above the transverse ridge, the rough surface of the occipital plane slopes forward, at a less obtuse angle with the basal plane, to the first named ridge which separates the occipital from the coronal or superior surface of the skull. The contour of this surface runs forwards, as far as the fragment extends, in an almost straight line: the extent of surface between the temporal muscular ridges must have been about five inches posteriorly, but it decreases gradually as it extends forwards: all that part which is preserved is quite smooth. The attachment of the fasciculi of the temporal muscle, and the convergence of their fibres as they passed through the zygoma are well marked on the sculptured surface of the bone. The zygomatic process is relatively stouter than in *Orycteropus:* it is prismatic: the external facet is nearly plane: the superior is concave, and increases in breadth anteriorly: the inferior surface offers a slight convexity behind the flattened articular surface for the lower jaw. The margin of the zygoma formed by the meeting of the upper and lower facets presents a semicircular curve, extended transversely from the cranium, and directed forwards.

The anterior extremity is obliquely truncated from below upwards and forwards, and presents a flattened triangular surface indicative of its junction with an os malæ: the space between this extremity and the side of the cranium measures one inch and nine lines across, and thus gives us the thickness of the temporal muscle. The distance from the origin of the zygoma to the occipital plane is relatively greater than in *Orycteropus; Glossotherium* is in this respect more similar to

Myrmecophaga and *Manis.*

The sphenoid bone forms a somewhat smooth protuberance below and behind the base of the *zygoma*. The tympanic bone is wedged in between this protuberance in front, and the mastoid process behind. The chief peculiarity of the broad mastoid is the regular semicircular cavity at its under part for the articulation of the styloid bone of the tongue. This depression is separated below by a broad rough protuberance from the foramen jugulare, (*f,* fig. 2, Pl. XVI,) which is immediately external to, and slightly in advance of the great foramen condyloideum, *c*. A small rugged portion of the os petrosum separates the jugular from the carotid canal, which arches upwards and directly inwards to the side of the shallow sella turcica, (the external and internal orifices of the carotid canal are shown at *g*, figs. 2 and 3). The chief protuberance on the basis cranii is a large and rugged one, serving for the attachment of muscles, and due chiefly to the expansion of a great sinus in the body of the sphenoid. This protuberance is separated from the smaller sphenoid protuberance before mentioned by a large groove continued downwards and forwards from the tympanic cavity, and containing the Eustachian tube, which does not traverse a complete osseous canal. Immediately internal to the glenoid cavity is the large orifice of the canal transmitting the third division of the fifth pair of nerves, the principal branch of which endows the tongue with sensibility; this foramen (*h*, fig. 2) is rather less than that for the muscular nerve of the tongue.

The internal surface of the present cranial fragment affords a very satisfactory idea of the size and shape of the brain of the extinct species to which it belongs. It is evident that, as in other Bruta, the cerebellum must have been almost entirely exposed behind the cerebrum; and that the latter was of small relative size, not exceeding that of the Ass; and chiefly remarkable, as in the Orycterope, Ant-eater, and Armadillo for the great development of the olfactory ganglia. The antero-posterior extent of the cribriform plate, as exposed in this fragment, is three inches, and the complication of the œthmoid olfactory lamellæ which radiate from it into the nasal cavity is equal to that which exists in the smaller Edentata (fig. 3, Pl. XVI). The nasal cavity is complicated in *Glossotherium* by the great number and capacious size of the air-cells which are in communication with it: these extend over all the upper, lateral, and back parts of the cranial cavity, as far even as the upper boundary of the foramen magnum: they also occupy the anterior two-thirds of the basis cranii. The external configuration of the skull would, therefore, afford a very inadequate or rather deceptive notion of the capacity of the cerebral cavity, were not the existence and magnitude of these sinuses known. The interspace of the outer and inner tables of the cranium are separated above the origins of the olfactory ganglia for the extent of three inches: above the middle of the cerebrum they are an inch and a half apart; at the sides of the cranium the interposed air-cells are from one to two inches across; at the back part of the cranium about one inch. The sinuses have generally a rounded form.

The foramen rotundum, (through which in figure 3 a probe is represented as passing), and the foramen ovale are situated close together, within a common transversely oblong depression (*i*). The carotid canal (*g*) opens into the outer side of the commencement of this wide channel, which conducts the great fifth pair of nerves to the outlets of its two chief divisions.

The petrous bone projects into the cranial cavity, in the form of an angular process with three facets: the foramen auditorium internum (*k*), and the aqueductus vestibuli, are situated on the posterior facet. Immediately behind the os petrosum is the foramen lacerum jugulare (*l*), situated at the point of convergence of the vertical groove of the lateral sinus, with a groove of similar size continued forwards from above the anterior condyloid canal. The plane of the internal opening of this canal (*c*, fig. 3) is directed obliquely inwards and backwards, and the lateral wall of the foramen magnum behind the foramen condyloideum slopes outwards to the edge of the condyle. Immediately internal to the foramen condyloideum is a small vascular foramen conducting a branch of the basilar artery into the condyloid canal, for the nourishment, doubtless, of the great lingual nerve.

In the relations of the plane of the internal orifice of the anterior condyloid foramen with that of the foramen magnum, we search in vain for a corresponding structure in any of the Mammiferous orders, save the Edentata:* and among these the Orycterope comes nearest the Glossothere in this respect. In the degree of development of the internal osseous ridge giving attachment to the tentorium cerebelli, the Ant-eaters and Armadillos more resemble the Glossothere than does the Orycterope; in which a continuous bony plate arches across the cranial cavity: in the Manis a still greater proportion of the tentorium is ossified, and it consequently recedes the furthest amongst the Edentata, in this, as in most other particulars of the cranial organization, from the Glossothere. The chief distinctive peculiarity in the cranium of the Glossothere, so far as it can be studied in the present fragment, and compared with that of other Edentata, is the deep, well-marked, semicircular styloid depression, above described.

A question may arise after perusing the preceding evidence, upon which the present fossil is referred to a great Edentate species nearly allied to the *Orycteropus*, whether one or other of the lower jaws, subsequently to be described, and in like manner referable, from their dentition, either to the *Orycteropodoid* or *Dasypodoid* families of Edentata, may not have belonged to the same species as does the present mutilated cranium. I can only answer, that those jaws were discovered by Mr. Darwin in a different and very remote locality,—that no fragments or teeth referable to them were found associated with the present fossil; and that, as it would be, therefore, impossible to determine from the evidence we have now before us, which of the two lower jaws should be associated with *Glossotherium;* and as

---

* In the monotrematous Echidna, the large canal for the lingual nerve has a widely different direction and course from that in the placental Edentata.

both may with equal if not greater probability belong to a totally distinct genus, it appears to me to be preferable, both in regard to the advancement of our knowledge of these most interesting Edentata of an ancient world, as well as for the convenience of their description, to assign to them, for the present, distinct generic appellations.

The figures in Plate XVI. preclude the necessity of a table of admeasurements of the cranial fragment of *Glossotherium*.

DESCRIPTION OF A MUTILATED LOWER JAW AND TEETH, ON WHICH IS FOUNDED A SUBGENUS OF MEGATHERIOID EDENTATA, UNDER THE NAME OF

MYLODON.

THE genus *Megalonyx*, as is well known, owes its name and the discovery of the fossil remains on which it was founded, to the celebrated Jefferson,* formerly President of the United States. Cuvier, from an examination of a single tooth, and the casts of certain bones of the extremities, especially the terminal ones, determined the ordinal affinities of this remarkable extinct quadruped.† But while he retained the name of *Megalonyx*, and used it in a generic sense, Cuvier offered no characters whereby other fossil remains might be generically either distinguished from, or identified with the *Megalonyx Jeffersonii*, unless, among such remains there happened to be a tooth, or a claw exactly corresponding with the descriptions and figures in the *Ossemens*

---

* Transactions of the Philosophical Society of Philadelphia, vol. iv. p. 246.

† Its relations to the Edentata, previously conjectured by Dr. Wistar, are proved in the Annales du Muséum, tom. v. p. 358; its more immediate affinities as an annectant form in that group are discussed in the edition of the Ossem. Fossiles, of 1833, tom. v. pt. 1. p. 160.

Fossiles; and when, of course, a specific identity, and not merely a generic relationship would be established.

The greater part of Cuvier's chapter on *Megalonyx* is devoted to the beautiful and justly celebrated reasoning on the ungueal phalanx, whereby it is proved to belong, not to a gigantic Carnivore of the Lion-kind, as Jefferson supposed, but to the less formidable order of Edentate quadrupeds; and Cuvier, in reference to the tooth,—the part on which alone a generic character could have been founded,—merely observes that it resembles at least as much the teeth of one of the great Armadillos, as it does those of the Sloths.‡

In the last edition of the *Règne Animal*, Cuvier introduces the *Megatherium* and *Megalonyx*, between the Sloths and Armadillos; but alludes to no other difference between the two genera than that of size,—"l'autre, le *Megalonyx*, est un peu moindre." (p. 226.) Some systematic naturalists, as Desmarest, and Fischer, have, therefore, suppressed the genus, and made the *Megalonyx* a species of *Megatherium* under the name of *Megatherium Jeffersonii*. The dental characters of the genus *Megatherium* are laid down by Fischer§ as follows: —"*Dent. prim. et lan. 0/0. molares 4/4-4/4, obducti, tritores, coronide nunc planâ transversim sulcatâ nunc medio excavatâ marginibus prominulis.*" That *Megalonyx* had the same number of molares as *Megatherium*, (supposing that number in the Megathere to be correctly stated, which it is not,) is here assumed from analogy, for neither Jefferson, Wistar, nor Cuvier, — the authorities for *Megalonyx* quoted by Fischer — possessed other means of knowing the dentition of that animal than were afforded by the fragment of a single tooth.

Now the almost entire lower jaw about to be described offers, in

---

‡ Speaking of this tooth, Cuvier observes, "Je l'avois cru d'abord nécessairement de paresseux; mais aujourd'hui que je connois mieux l'ostéologie des divers tatous, je trouve qu'elle ressemble au moins autant à une dent de l'un des grands tatous.— Loc. cit. p. 172.

§ Synopsis Mammalium.

so far as respects the general form and structure of the teeth, the same kind and degree of correspondence with the *Megatherium*, as does the *Megalonyx Jeffersonii* of Cuvier: and, what is only probable in that species, is here certain, viz., an agreement with the Megatherium in the class, viz. *molares*, to which the teeth exclusively belong. The question, therefore, on which I find myself, in the outset, called upon to come to a decision is, as to the preference of the mode of viewing the subject of the generic relationship of the *Megalonyx* adopted by Desmarest, Fischer, &c., or of that, on which Cuvier, and after him Dr. Harlan, have practically acted: whether, in short, the genus *Megatherium* is to rest upon the more comprehensive characters of kind and general structure of the teeth, or upon the more restricted ones, of form and such modifications in the disposition and proportions of the component textures of the tooth, as give rise to the characteristic appearances of the triturating surface of the crown.

With respect to existing Mammalia, most naturalists of the present day seem to be unanimous as to the convenience at least of founding a generic or sub-generic distinction on well marked modifications in the form and structure of the teeth, although they may correspond in number and kind, in proof of which it needs only to peruse the pages of a *Systema Mammalium* which relate to the distribution of the Rodent Order. According to this mode of viewing the logical abstractions under which species are grouped together, the extinct Edentate Mammal discovered by Jefferson must be referred to a genus distinct from *Megatherium*, and for which the term *Megalonyx* should be retained. This will be sufficiently evident by comparing the descriptions given by Cuvier of one of the teeth of the *Megalonyx Jeffersonii*, and by Dr. Harlan of a tooth of his *Megalonyx laqueatus*, with those of the *Megatherium* which have been published by Mr. Clift. The fragment of the molar tooth of the *Megalonyx Jeffersonii*, described and figured in the *Ossemens Fossiles*, seems to have been implanted in the jaw, like the teeth of the *Megatherium*, by a simple hollow base similar in form and size to the protruded crown: its structure Cuvier describes as consisting of a central cylinder of bone enveloped in a sheath of enamel.* The transverse section of this tooth presents an irregular elliptical form, the external contour being gently and uniformly convex, the internal one, undulating; convex in the middle, and slightly concave on each side, arising from the tooth being traversed longitudinally on its inner side by two wide and shallow depressions.

The imperfect tooth of the species called by Dr. Harlan *Megalonyx laqueatus*, and of which a cast was presented by that able and industrious naturalist to the Museum of the Royal College of Surgeons, resembles in general form, and especially in the characteristic double longitudinal groove on the inner side, the tooth of the *Megalonyx Jeffersonii*. It is thus described by Dr. Harlan:

"The fractured molar tooth appears to have belonged to the inferior maxilla on the right side; the crown is destroyed; a part of the cavity of the root remains. The body is compressed transversely, and presents a double curvature, which renders its anterior and exterior aspects slightly convex; the posterior and interior gently concave; these surfaces are all uniform, with the exception of the interior or mesial aspect, which presents a longitudinal rib or ridge, one-half the thickness of the long diameter of the tooth; with a broad, not profound longitudinal groove or channel along each of its borders. It is from this resemblance to a portion of a fluted column, that the animal takes its specific appellation (*Megx. laqueatus*).

"The crown would resemble an irregular ellipsis widest at the anterior portion. The tooth consists of a central pillar of bone surrounded with enamel, the former of a dead white, the latter of a ferruginous brown colour: the transverse diameter is more than two-thirds less than its length, whilst that of *Meg^x. Jeffersonii* is

---

*   It is most probable that the substance which is here termed " enamel," is similar to that which forms the dense prominent ridges in the tooth of the Megatherium, and which I have shown to be composed of minute parallel calcigerous tubes, similar to the ivory or bone of the human tooth.

only one-third less—the antero-posterior diameter is one-half its length in the former, and two-thirds less in the latter. The proportions of this tooth are consequently totally at variance with that of its kindred species." [Vide Pl. XII. fig. 7, 8, 9.]*

Dr. Harlan describes also two claws of the fore-foot, a radius, humerus, scapula, one rib, an os calcis, a metacarpal bone, certain vertebræ, a femur, and tibia, of the same *Megalonyx;* these parts of the skeleton, together with the tooth, which so fortunately served to establish the generic relationship of the species with the *Megalonyx* of Jefferson and Cuvier, were discovered in Big-bone-cave, Tenessee, United States.

Dr. Harlan does not enter into the question of the generic characters of *Megalonyx*, but it would seem that he felt them to rest not entirely on dental modifications, for he observes that "a minute examination of the tooth and knee-joint renders it not improbable, supposing the last named character to be peculiar to it, that if the whole frame should hereafter be discovered, it may even claim a generic distinction, in which case, either Aulaxodon, or PLEURODON, would not be an inappropriate name."†

There can be no doubt, as it appears to me, with respect to a fossil jaw presenting teeth in the same number, and of the same general structure, as in the *Megatherium*, and with individual modifications of form, as well marked as those which distinguish *Megatherium* from *Megalonyx*, that the Palæontologist has no other choice than to refer it, either as Fischer has done with *Megalonyx*, to a distinct species of the genus *Megatherium*, or to regard it as the type of a subgenus distinct from both. With reference, however, to the *Pleurodon* of Dr. Harlan, after a detailed comparison of the cast of the tooth on which that genus is mainly founded, with the descriptions and figures of the tooth of the *Megalonyx Jeffersonii*, in the "Ossemens Fossiles," they seem to differ in so slight a degree as to warrant only a

---

* Medical and Physical Researches, pp. 323-4.
† Loc cit. p. 330.

specific distinction, and this difference even, viewing the various proportions of the teeth in the same jaw of the *Megatherium*, is more satisfactorily established by the characters pointed out by Dr. Harlan in the form and proportions of the radius, than by those in the tooth itself.

The next notice of the *Megalonyx* which I have consulted, in the hope of meeting with additional and more precise information as to its real generic characters, is an account given by the learned Professor Doellinger,‡ of some fossil bones, collected by the accomplished travellers Spix and Martius in the cave of Lassa Grande, near the Arrayal de Torracigos, in Brazil. In this collection, however, it unfortunately happens that there are no teeth, but only a few bones of the extremities, including some ungueal phalanges, which Professor Doellinger concludes, from their shape, the presence of an osseous sheath for the claw, and the form of their articulation, to belong, without doubt, to an animal of the Megatherioid kind, about the size of an Ox. He particularly states that they are not bones of an immature individual; but that they agree sufficiently with Cuvier's descriptions and figures of the *Megalonyx* to be referred to that species of animal (zu dieses thierart;) and he adds, what is certainly an interesting fact, that the fossils in question form the first of the kind that had been discovered out of North America.

Subsequently to the discovery of these bones, and of those of the *Megalonyx laqueatus* above alluded to, the remains of another great Edentate animal were found in North America, and were deposited in the Lyceum at New York; among these is a portion of the lower jaw with the whole dental series of one side. It is thus described by Dr. Harlan.

"The fragment I am now about to describe is a portion of the dexter lower jaw of the Megalonyx, containing four molar teeth; three of the crowns of these teeth are perfect, that of the anterior one is imperfect. These teeth differ considerably from each

---

‡ Spix and Martius, Reise in Brazil, Band ii. p. 5.

other in shape, and increase in size from the front, the fourth and posterior tooth being double the size of the first, and more compressed laterally; it is also vertically concave on its external aspect, and vertically convex on its internal aspect; the interior or mesial surface is strongly fluted, and it has a deep longitudinal furrow on the dermal aspect, in which respect it differs from the tooth of the *M. laqueatus* previously described by me, of which the dermal aspect is uniform, but to which, in all other respects, it has a close resemblance. I suppose it therefore probable, that this last may have belonged to the upper jaw. The three anterior molars differ in shape and markings: they are vertically grooved, or fluted, on their interior and posterior aspects, a transverse section presenting an irregular cube. The length of the crown of the posterior molar is two inches: the breadth about five-tenths of an inch: the length of the tooth is three inches and six-tenths. The diameter of the penultimate molar is eight-tenths by seven-tenths of an inch. The length of this fragment of the jaw-bone is eight inches and four-tenths; the height three inches and six-tenths: the length of the space occupied by the alveolar sockets five inches and eight-tenths. The crown of the tooth presents no protuberances, but resembles that of the Sloth; the roots are hollow."*

This fossil is referred by Dr. Harlan to his *Megalonyx laqueatus;* but, pending the absence of other proof of the identity of species, in which, as may be seen by comparing fig. 2, with fig. 4, in Pl. XVII., the teeth differ widely in form, it would be obviously

---

* Harlan's Medical and Physical Researches, 1835, p. 334. M. de Blainville speaks of a cast of a fragment of a lower jaw "portant encore *cinq* dents en série;" as having been transmitted to the Museum of the Garden of Plants from North America, together with other bones, all of which he refers to the genus *Megalonyx*; M. de Blainville does not describe these teeth, which is to be regretted, inasmuch as, if he be correct in regard to their number, which can hardly be doubted, and if he wrote with any clear and definite ideas of the generic characters of *Megalonyx*, this would indicate that *Megalonyx* differed generically both from *Megatherium* and *Mylodon* in a more important dental character than has hitherto been suspected (See "Comptes Rendus, &c." 1839, No. V. p. 142.)

hazardous to adopt such an approximation on hypothetical grounds.[†] In order, however, to obtain more satisfactory evidence of the nature and amount of the difference between the *Megalonyx laqueatus*, and the allied animal represented by the above-described fragment of lower jaw, I wrote to my much respected friend M. LAURILLARD, requesting him to send me a sketch of the teeth in the cast of that lower jaw, which had been transmitted from New York to the Garden of Plants. With full confidence in the characteristic precision and accuracy of the drawing with which I have been obligingly favoured by M. Laurillard, I am disposed to regard the amount of difference recognizable in every tooth in the lower jaw in question (fig, 3 and 4,) as compared with the molar tooth either of *Megalonyx Jeffersonii* (fig. 1,) or *Megx. laqueatus* (fig. 2) to be such as to justify its generic separation from *Megalonyx* on the same grounds as *Megalonyx* is distinguished from *Megatherium*, and for the subgenus of Megatherioid Edentata, thus indicated, I would propose the name of MYLODON.[‡] The species of which the fossil remains are described by Dr. Harlan may be dedicated to that indefatigable Naturalist who has contributed to natural science so much valuable information respecting the Zoology, both recent and fossil, of the North American continent. The fossil about to be described represents a second and smaller species of the same genus, and I propose to call it *Mylodon Darwinii*, in honour of its discoverer, of whose researches in the Southern division of the New World it forms one of many new

---

[†] Dr. Harlan also indicates differences in certain parts of the skeleton of the New York fossils as compared with his *Megx. laqueatus*; but thinks them probably due to a difference in the age of the individuals: he says "There is also in Mr. Graves' collection, in New York, a tibia, nearly perfect from the right leg; the segment of a flattened sphere, on which the external condyle of the femur moves, is rather more depressed, than in the specimen from Big-bone-cave. Other marks and peculiarities are observable on this bone, not found on that of the *Megalonyx laqueatus* of Big-bone-cave, but they are probably due to a difference in the age of the individuals." Loc. cit. p. 335.

[‡] Μυλη, *mola*; οδους, *dens*.

and interesting fruits.

This fossil was discovered in a bed of partly consolidated gravel at the base of the cliff called Punta Alta, at Bahia Blanca in Northern Patagonia: it consists of the lower jaw with the series of teeth entire on both sides: but the extremity of the symphysis, the coronoid and condyloid processes, and the angular process of the left ramus, are wanting. The teeth are composed, as in *Bradypus*, *Megatherium* and *Megalonyx*, of a central pillar of coarse ivory, immediately invested with a thin layer of fine and dense ivory, and the whole surrounded by a thick coating of cement.

In the fig. 5, Pl. XVII., the fine ivory is represented by the white striated concentric tract on the grinding surface of the teeth; it is of a yellowish-white colour in the fossil, and stands out, as an obtuse ridge, from that surface: both these conditions depend on the large proportion of the mineral to the animal constituent in this substance of the tooth. The external layer of the cement presents in the fossil the same yellowish-brown tint as the bone itself, which it so closely resembles, both in intimate structure and in chemical composition; the internal layer next the dense ivory is jet black, indicating the great proportion of animal matter originally present in this part. The central pillar of coarse ivory, which, from its more yielding texture, has been worn down into a hollow at the triturating surface of the tooth, also presents, as a consequence of the less proportion of the hardening phosphates, a darker brown colour than the external layer of the cement, or the bone itself.

The teeth are implanted in very deep sockets; about one-sixth only of the last molar projects above the alveolus; the proportion of the exposed part of the tooth increases as they are placed further forwards. The implanted part of each tooth is simple; preserving the same size and form as the projecting crown, and presenting a large conical cavity at the base, indicative of the original persistent pulp, and perpetual growth of these teeth.

The extent of the whole four alveoli is four inches, eight lines; the length of the jaw from the angle to the broken end of the symphysis is seventeen inches and a half;* from the figures it will be seen that only a small proportion of the anterior part of the jaw is lost, so that we may regard the dentigerous part of the jaw as being limited to about one-fourth of its entire length; the alveoli being nearly equidistant from the two extremities. The first and second teeth, counting backwards, are separated by an interspace of rather more than three lines; that between the second and third is one line less; the third and fourth are rather more than a line apart: from the oblique position, however, of the three hinder teeth the intervals between them appear in a side view, as in fig. 1, Pl. XIX., to be less than in reality, and the third and fourth teeth seem to touch each other.

Each tooth has a form and size peculiar to itself, and different from the rest, but corresponds of course with its fellow on the opposite side. The same may be observed, but in a less degree, in the teeth of the Megatherium itself; hence, it is obviously hazardous to found a generic distinction upon a single tooth, unless, as in the case of the *Glyptodon*,† the modification of form happens to be extremely well marked. The whole series of teeth, or their sockets, at least of one of the jaws, should be known for the purpose of making a satisfactory comparison with the previously established Edentate genera.

The first molar in the present jaw is the smallest and simplest of the series: its transverse section is ellipsoid, or subovate, narrowest in front, and somewhat more convex on the outer than on the inner side: the long diameter of the ellipse is nine lines, the short or transverse diameter six lines: the length of the tooth may be about three inches, but I have not deemed it necessary to fracture the alveolus in order to ascertain precisely this point.

The second tooth presents in transverse section a more irregular and wider oval figure than the first: the line of the

---

* If the lower jaw of *Mylodon Harlani*, bears the same proportion to its teeth as does that of *Mylodon Darwinii*, it must be about two feet in length.

† See Proceedings of the Geological Society, March 1839, and Parish's Buenos Ayres, p. 178, *b*, Pl. 1, fig. 2 and 3.

outer side is convex, but that of the inner side slightly concave, in consequence of the tooth being traversed longitudinally by a broad and shallow channel or impression; the longitudinal diameter of the transverse section is one inch; the transverse diameter at the widest part nine lines. There is a slight difference in the size of this tooth on the two sides of the jaw, the right one, from which the above dimensions are taken, being the largest.

The transverse section of the third tooth has a trapezoidal or rhomboidal form; the angles are rounded off; the posterior one is most produced; the anterior and posterior surfaces are flattened, the latter slightly concave in the middle; the external and internal sides are concave in the middle, especially the inner side, where the concavity approaches to the form of an entering notch. The longest diameter of the transverse section of this tooth is thirteen lines, the shortest seven lines and a half: in the tooth on the right side the external surface is nearly flat; this slight difference is not indicated in the figure (Pl. XVIII.)

The last molar, which is generally the most characteristic in the fossil *Bruta*, presents in an exaggerated degree the peculiarities of the preceding tooth; the longitudinal channels on both the outer and inner surfaces encroach so far upon the substance of the tooth, that the central coarse ivory substance is as it were squeezed out of the interspace, and the elevated ridge of the dense ivory describes an hour-glass figure upon the triturating surface, the connecting isthmus being but half the breadth of the rest of the tract; the external cæmentum preserves nearly an equal thickness throughout. Of the two lobes into which this tooth is divided by the transverse constriction, the anterior is the largest; their proportions and oblique position are pretty accurately given in the figure. The longitudinal diameter of the transverse section of this tooth is one inch, seven lines, its greatest lateral or transverse diameter is ten lines, its least diameter at the constricted part is three lines, the length of the entire tooth is four inches. Judging from the form of the jaw, the length of the other teeth decreases in a regular ratio to the anterior one. The posterior tooth is slightly curved, as shown in fig. 2, Pl. XIX., with the concavity directed towards the outer side of the jaw.

The general form of the horizontal ramus of the jaw, is so well illustrated in the figures Pl. XVIII. and XIX., that the description may be brief.

The symphysis is completely anchylosed, about four inches in length, and extended forward to the extremity of the jaw at a very slight angle with the inferior border of the ramus: it is of great breadth, smooth and gently concave internally, and suggests the idea of its adaptation for the support and gliding movements forwards and backwards of the free extremity of a long and well-developed tongue.

The exterior surface of the symphysis is characterized by the presence of two oval mammilloid processes, situated on each side of the middle-line, and about half way between the anterior and posterior extremes of the symphysis. A front view of these processes, of the natural size, is given in fig. 4, Pl. XIX.: a side view of the one on the right side represented in the reduced figure.

Nearly four inches behind the anterior extremity of the above process is the large anterior opening of the dental canal: it is five lines in diameter, situated about one-third of the depth of the ramus of the jaw from the upper margin. The magnitude of this foramen, which gives passage to the nerve and artery of the lower lip, indicates that this part was of large size; and the two symphyseal processes, which probably were subservient to the attachment of large retractor muscles, denote the free and extensive motions of such a lip, as we have presumed to have existed from the size of the foramina destined for the transmission of its nervous and nutrient organs.

The angle of the jaw is produced backwards, and ends in an obtuse point, slightly bent upwards; a foramen, one-third less than the anterior one, leads from near the commencement of the dental canal, to the outer surface of the jaw, a little below and behind the last molar tooth; this foramen presents the same size and relative position on both sides of the jaw. I find no indication of a corresponding foramen, or of symphyseal processes in the figures or descriptions of the lower jaw of the Megatherium, nor in the lower jaw of the Sloths, Ant-eaters, Armadillos, or Manises, which I have had the opportunity of examining with a view to this comparison.

In the Megatherium the inferior contour of the lower jaw is peculiarly remarkable, as Cuvier has observed, for the convex prominence or enlargement which is developed downwards from its middle part. In the Mylodon the corresponding convexity exists in a very slight degree, not exceeding that which may be observed at the corresponding part of the lower jaw of the Ai, or Orycterope. A broad and shallow furrow extends along the outer side of the jaw, close to the alveolar margin, from the beginning of the coronoid process to the anterior dental foramen.

The base of the coronoid process begins external and posterior to the last grinder: the whole of the ascending ramus of the jaw, beneath the coronoid process is excavated on its inner side by a wide and deep concavity, bounded below by a well-marked ridge, which extends obliquely backwards from the posterior part of the alveolus of the last grinder to the inferior margin of the ascending ramus, which is bent inwards before it reaches the angle of the jaw.

The large foramen or entry to the dental canal is situated in the internal concavity of the ascending ramus of the jaw, two inches behind the last molar, three inches from the lower margin of the ramus, and nearly five inches from the elevated angle of the jaw: it measures nine lines in the vertical diameter, and its magnitude indicates the large size of the vessels which are destined to supply the materials for the constant renewal of the dental substance, — a substance which from its texture must be supposed to have been subject to rapid abrasion. About an inch behind the dental foramen a deep vascular groove, about two lines in breadth, is continued downwards to the ridge which circumscribes the internal concavity of this part of the jaw, and perforates the ridge, which thus arches over the canal: this structure is present in both rami of the jaw. The mylo-hyoid ridge is distinctly marked about an inch and a half below the alveolar margin. Other muscular ridges and irregular eminences are present on the outer side of the base of the ascending ramus, and near the angle of the jaw; as shown in fig. 1, Pl. XIX.

From the preceding descriptions it will be seen that the lower jaw of the *Mylodon* is very different from that of the *Megatherium*; with that of the *Megalonyx* we have at present no means of comparing it. Among existing Edentata the Mylodon, in the form of the posterior part and angle of the jaw, holds an intermediate place between the Ai and the great Armadillo; in the form of the anchylosed symphysis of the lower jaw it resembles most closely the Unau or two-toed Sloth; but in the peculiar external configuration of the symphysis resulting from the mammilloid processes above described, the Mylodon presents a character which has not hitherto been observed in any other species of *Bruta*, either recent or fossil.

In conclusion it may be stated, that the teeth and bones here described offer all the conditions and appearances of those of a full grown animal; and that they present a marked difference of size as compared with those of the *Mylodon Harlani*, as will be evident by the following admeasurements.

ADMEASUREMENTS OF THE LOWER JAW OF MYLODON DARWINII.

	Inches.	Lines.
Length (as far as complete)	17	6
Extreme width, from the outside of one ramus to that of the other ..	9	0
Depth of each ramus	4	9
Length of alveolar series	4	8
From first molar to broken end of symphysis	6	0
Breadth of symphysis	3	7
Longitudinal extent of symphysis	4	6
Circumference of narrowest part of each ramus	5	9

DESCRIPTION OF A CONSIDERABLE PART OF THE SKELETON OF A LARGE EDENTATE MAMMAL, ALLIED TO THE MEGATHERIUM AND ORYCTEROPUS, AND FOR WHICH IS PROPOSED THE NAME OF

SCELIDOTHERIUM* LEPTOCEPHALUM.

OF the large Edentate quadrupeds that once existed in the New World, sufficient of the osseous remains of the gigantic Megatherium alone has been transmitted to Europe to give a satisfactory idea of the general form and proportions of the extinct animal.

Different bones of the Megalonyx, Mylodon, and Glyptodon have been described, but not sufficient of the remains of any

---

* Σκελις, *femur*; θηριον, *bellua*; in allusion to the disproportionate size of the thigh-bone.

individual of these subgenera has, hitherto, reached Europe, or been so described as to enable us to form a comparison between them and the Megatherium, or any of the existing Edentata, in regard to the general construction and proportions of the entire skeleton.

This state of our knowledge of the osteology of the singular giants of the Edentate Order renders the remains of the present animal peculiarly interesting, since, although the extremities are too imperfect to enable us to reconstruct the entire skeleton, a sufficient proportion of it has been preserved in the natural position to give a very satisfactory idea of its affinities to other Edentata, whose osteology is more completely known.

The fossil remains here described were discovered by Mr. Darwin in the same bed of partly consolidated gravel at Punta Alta, Northern Patagonia, as that in which the lower jaws of the *Toxodon* and *Mylodon* were imbedded. The parts of the skeleton about to be described were discovered in their natural relative position, as represented at Pl. XX., indicating, Mr. Darwin observes, that the sublittoral formation in which they had been originally deposited had been subject to little disturbance.† They include the cranium, nearly entire, with the teeth and part of the os hyoides; the seven cervical, eight of the dorsal, and five of the sacral vertebræ, the two scapulæ, left humerus, radius and ulna, two carpal bones, and an ungueal phalanx; both femora, the proximal extremities of the left tibia and fibula, and the left astragalus.

The principal parts of the cranium which are deficient are the anterior extremities of both the upper and lower jaws, the os frontis, æthmoid bone, and the whole upper part of the facial division of the skull; but sufficient remains to show that the general form of the skull resembled an elongated, slender, sub-compressed cone, commencing behind by a flattened

---

† This beach is covered at spring tides; many parts of the skeleton were encrusted with recent *Flustræ*, and small marine shells were lodged in the crevices between the bones.

vertical base, slightly expanding to the zygomatic region, and thence gradually contracting in all its dimensions to the anterior extremity.

The Cape Ant-eater (*Orycteropus*), of all Edentata, most nearly resembles the present fossil in the form of its cranium, and next in this comparison the great Armadillo (*Dasypus gigas*, Cuv.) may be cited: on the supposition, therefore, that the correspondence with the above existing Edentals observable in the parts of the fossil cranium which do exist, was carried out through those which are defective, the length of the skull of the Scelidothere must have been not less than two feet. If now the reader will turn to Pl. XX. he will see that this cranium is singularly small and slender in proportion to the rest of the skeleton, especially the bulky pelvis and femur, of which bones the latter has a length of seventeen inches, and a breadth of not less than nine inches; the astragalus, again, exceeds in bulk that of the largest Hippopotamus or Rhinoceros; yet the condition of the epiphyseal extremities of the long bones proves the present fossils to have belonged to an immature animal. Hence, although the Scelidothere, like most other Edentals, was of low stature, and, like the Megatherium, presented a disproportionate development of the hinder parts, it is probable, that, bulk for bulk, it equalled, when alive, the largest existing pachyderms, not proboscidian. There is no evidence that it possessed a tesselated osseous coat of mail.

I shall commence the description of the present skeleton with the cranium.

The condyles of the occiput (See Pl. XXI. fig. 2,) are wide apart, sub-elliptic, very similar in position, form, and relative size to those in *Orycteropus*. The foramen occipitale is transversely oval, its plane slopes from above downwards and forwards at an angle of 40° with that of the occipital region of the skull. This region, as before stated, is vertical in position (see fig. 1, Pl. XXI.), of a sub-semicircular form, the breadth being nearly one-third more than the height; it is bounded above and laterally by a pretty regular curve; but the superior margin is not produced so far backwards as in *Orycteropus*. The occipital plane is bisected by a mesial vertical ridge; there is a less developed transverse curved intermuscular crest which runs parallel with and about half an inch below the marginal ridge: the surface of the occipital plane on the interspaces of these ridges is irregularly pitted with the impression of the insertion of powerful muscles. The corresponding surface is smooth in the Orycterope and Armadillos; in the great extinct Glossothere it resembles in character that of the Scelidothere; but in the forward slope of the occipital plane the Glossothere differs in a marked degree from the present animal.

The upper surface of the cranium is smooth and regularly convex. The extent of the origin of the temporal muscles is defined by a slightly-raised broad commencement of a ridge, which, in the older animal, might become more developed. There is no trace of this ridge in the Orycterope; but it exists in the Armadillos, in which the teeth are of a denser texture, and better organized for mastication, and consequently are associated with better developed masticatory muscles. It will be subsequently shown that the Scelidothere resembles the Armadillos in so far as it possesses a greater proportion of the dense ivory to the external cæmentum in its teeth, than does the Megatherium; while it differs widely from the Orycterope, in the structure of its teeth. The teeth, however, are fewer in the Scelidothere than in any Armadillo, and relatively smaller than in most of the species of that family. Accordingly we find that the zygomatic arches are relatively weaker; and in this particular the *Scelidothere* corresponds with the Orycterope. The zygomatic process of the temporal commences posteriorly about an inch and a half from the occipital plane, its origin or base is extended forwards in a horizontal line fully four inches, where it terminates as usual in a thin concave edge, as shown on the right side in Pl. XXII. The free portion of the zygoma, continued forwards from the outer part of this edge, is a slender sub-compressed process, half an inch in the longest or vertical diameter, and less than three lines in the transverse; the extremity of this process is broken off; the opposite extremity of the malar portion of the zygoma is entire, and obtusely rounded. The bony arch may have been completed by the extension of the temporal process to the malar one, but the two parts undoubtedly were not connected together by so extensive a surface as in the Orycterope. On the other hand, if the zygomatic arch be naturally incomplete in the Scelidothere, the interspace between the malar and temporal portions must be relatively much less than in the Sloth or Ant-eater; for the broken end of the temporal part is separated from the obtusely rounded apex of the malar process in the present specimen by an interval of only one inch.

The articular surface (Pl. XXIII., fig. 2) beneath the zygoma for the lower jaw is flat and even, with the outer and inner margin slightly bent down, but having no definable anterior or posterior limits; its breadth is two inches. It differs from the corresponding surface in the Orycterope in being separated by a relatively wider interval from the tympanic bone, and in wanting consequently the support which the bony meatus auditorius gives in the Orycterope to the back part of the mandibular joint. The Armadillos differ still more from the Scelidothere in this important part of the cranial organization, inasmuch as the glenoid cavity is not only protected behind by the descending os tympanicum, but also in front by a corresponding vertical downward extension of the os malæ. The Scelidothere in the general form and relative position of the surface for the articulation of the lower jaw resembles the Glossothere more closely than any other Edentate animal with which I have been able to compare it.

The malar bone of the Megatherium presents, as is well known, two characters, in which it conspicuously differs from that of the Orycterope and Armadillos, and approximates in an equally marked degree to the Sloths; these characters consist in a process ascending as if to complete the posterior circumference of the orbit, and another process descending outside the lower jaw to give advantageous and augmented surface of attachment to the masseteric muscle, in its character of a protractor of the jaw. Now both these modifications of the malar bone are present in the Scelidothere, and are the chief if not the sole marks of the affinity to the Megatherium which the structure of the cranium affords. They are, however, the more interesting, perhaps, on that account, and because they are associated with other and more numerous characters approximating the species in question to the ordinary terrestrial as distinguished from the arboreal Edentata. For if the Scelidothere, instead of the Megathere, had been discovered half a century ago, and if its true nature and affinities had been in like manner elucidated by the genius and science of a Cuvier; and supposing on the other hand that the Megatherium instead of the Scelidothere had been one of the novel and interesting fruits of Mr. Darwin's recent exploration of the coast of South America, then the affinities of the Megathere with the Sloths would undoubtedly have been viewed from a truer point than at the time when,—the Scelidothere, and analogous transitional forms, being unknown,—it was regarded as a gigantic Sloth.

Having indicated the principal characters of the cranium of the Scelidothere, which determine its affinities amongst the *Edentata*, there next remains to be considered the relative position, extent, and connections, of the different bones composing the cranium.

The occipital bone constitutes the whole of the posterior, the usual proportion of the inferior, and a small part of the upper and lateral portions of the cranial cavity: there is a small descending ex-occipital process immediately exterior to the condyle: above this part the occipital bone is articulated to the mastoid process of the temporal, and the supra-occipital plate is joined by a complex dentated lambdoidal suture to the two parietals, without the intervention of interparietal or Wormian bones; the course and form of the lambdoidal suture is shown in Pl.XXII; it has the same relative position as in the Orycterope; in the Armadillos, the suture runs along the angle between the posterior and superior surfaces of the skull. The thickness of the occipital bone, at this angle, in the Scelidothere, exceeds an inch, and its texture consists of a close massive diploë, between the dense outer and inner tables, (Pl. XXIII. fig. 1.)

The squamous portion of the temporal bone has a very slight elevation, not extending upon the side of the cranium more than half an inch above the zygoma; it is thus relatively lower than in the *Orycteropus;* but is similarly bounded above by an almost straight line, (Pl. XXI., fig. 1.) The mastoid process is small, compressed, with a rounded contour; immediately internal to it is a very deep depression, corresponding to that for the digastric muscle. But the most interesting features in this region of the temporal bone consist in the free condition of the tympanic bones, and the presence of a semicircular pit, immediately behind the tympanic bone for the articulation of the styloid element of the hyoid or tongue-bone: in these points we trace a most remarkable correspondence with the Glossothere, and in the separate tympanic bone the same affinity to the Orycteropus, as has been already noticed in the more bulky extinct Edental.

This correspondence naturally leads to a speculation as to the probable generic relationship between the Glossothere and Scelidothere: now it may first be remarked that the styloid articular depression is relatively much larger and much deeper in the Glossothere than in the Scelidothere; in the former its diameter equals, as we have seen, one inch; in the Scelidothere it measures only a third of an inch, the whole cranium being about two-fifths smaller; if we turn next to the anterior condyloid foramina, which in the Scelidothere are double on each side, we obtain from them evidence that the muscular nerve of the tongue could only have been one-third the size of that of the Glossothere. These proofs of the superior relative development of the tongue in the Glossothere indicate a difference of habits, and a modification, probably, of the structure of the locomotive extremities; and when we associate these deviations from the Scelidothere, with the known difference in the position of the occipital plane, which in the Glossothere corresponds with that in the *Myrmecophaga* and *Bradypus*, we shall be justified in continuing to regard them, until evidence to the contrary be obtained, as belonging to distinct genera.

The parietal bones present an oblong regular quadrate figure, the sagittal suture running parallel with the squamous, and the frontal with the lambdoidal suture; there is scarcely any trace of denticulations in the sagittal suture; the bones are of remarkable thickness, varying, at this suture, from six to nine lines, and their opposed surfaces are locked together by narrow ridges, which slightly radiate from the lower to the upper part of the uniting surface: the substance of the bone consists of an uniform and pretty dense diploë; and there are no sinuses developed in it. We can hardly regard the extraordinary air-cells which occupy the interspace of the two tables of the skull in the parietal and occipital bones of the Glossothere (Pl. XVI., fig. 3) as a difference depending merely on age.

The frontal and æthmoid bones are broken away in the present cranium. The sphenoid commences two inches in front of the foramen occipitale; the fractured state of the skull does not allow its anterior or lateral limits to be accurately defined; its body is occupied with large air-sinuses; the only part, indeed, of this bone which is exposed to observation is that which forms part of the floor of the cranium; and this we shall now proceed to describe, in connexion with the other peculiarities of the cranial cavity, (fig. 1. Pl. XXIII.) The body of the sphenoid is impressed on its cranial surface with a broad and shallow sella turcica (*a*), bounded by two grooves, (*b b*,) leading forwards and inwards from the carotid foramina (*c*); the line of suture between the sphenoid and occipital bones runs along a slight transverse elevation (*d*), which bounds the sella posteriorly; this suture is partially obliterated: a slight median protuberance (*e*) bounds the sella turcica anteriorly; there are neither anterior nor posterior clinoid processes. External to the carotid channel there is a wide groove (*f*) leading to the foramen ovale (*g*); this foramen is about one-third smaller than in the Glossothere, and therefore, as compared with the anterior condyloid foramina, indicates that the tongue was endowed with a greater proportion of sensitive than motive power in the Scelidothere: but in reasoning on the size of this nerve, it must be remembered that in both animals certain branches, both of the second and third divisions of the fifth pair of nerves, are to be associated with the persistence of large dental pulps, of which they regulate the secreting power. Anterior to the foramen ovale, and at the termination of the same large common groove, lodging the trunk of the fifth pair of nerves is the foramen rotundum (*h*); this leads to a very long canal, the diameter of which is five lines, being somewhat less than that for the third division of the fifth pair. The anterior sphenoid is broken away, so that no observation can be made on the optic foramina.

The basilar process of the occipital bone is perforated at its middle by two small foramina (*i*) on the same transverse line, about half an inch apart.

In the Armadillo these foramina do not exist: in the Orycterope they are present, but open beneath an overhanging ridge, which is continued from them to the upper part of the anterior condyloid foramen on each side. The sella turcica of the Orycterope is deeper and narrower than in the Scelidothere; and is separated from the basilar occipital process by a transverse ridge, which sends forward two short clinoid processes; two smaller anterior clinoid processes project backwards from the angle of the anterior boundary of the sella turcica. The foramina ovalia and rotunda open in the same continuous groove, as in the Glossothere and Scelidothere, but they are relatively wider apart; and the canal for the third division of the fifth pair is shorter, and runs more directly outwards.

The petrous bone in the Scelidothere is relatively larger than in the Glossothere, but this probably arises from the precocious development of the organ of hearing in the present immature specimen in obedience to the general law. The trunk of the fifth pair of nerves does not impress it with so deep and well defined a groove as in the Glossothere; the elliptic internal auditory foramen (*k*) is situated about the middle of the posterior surface; behind this is the aqueductus vestibuli; and immediately posterior to the petrous bone is the foramen jugulare (*l*): the shape of the os petrosum agrees more with that of the Armadillo than with that of the Orycterope. An accidental fracture of the right os petrosum demonstrates its usual dense and brittle texture, and at the same time has exposed the cochlea with part of its delicate and beautiful lamina spiralis. The conservation of parts of the organs of vision in certain fossils, has given rise to arguments which prove that the laws of light were the same at remote epochs of the earth's history as now; and the structures I have just mentioned, in like manner, demonstrate that the laws of acoustics have not changed, and that the extinct giants of a former race of quadrupeds were endowed with the same exquisite mechanism for appreciating the vibrations of sound as their existing congeners enjoy at the present day.

The brain, being regulated in its development by laws analogous to those which govern the early perfection of the organ of hearing, appears to have been relatively larger in the Scelidothere than in the Glossothere: it was certainly relatively longer; the fractured cranium gives us six inches of the antero-posterior diameter of the brain, but the analogy of the Orycterope would lead to the inference that it extended further into the part which is broken away. The greatest transverse diameter of the cranial cavity is four inches eight lines: these dimensions, however, are sufficient to show that the brain was of very small relative size in the Scelidothere; and, both in this respect, and in the relative position of its principal masses, the brain of the extinct Edental closely accords with the general character of this organ in the existing species of the same Order. We perceive by the obtuse ridge continued obliquely upwards from above the upper edge of the petrous bone, that the cerebellum has been situated wholly behind the cerebrum; we learn also from the same structure of the enduring parts that these perishable masses were not divided, as in the Manis, by a bony septum, but by a membraneous tentorium, as in the Glossothere and Armadillos: in the Orycteropus, as has been before remarked, there is a strong, sharp, bony ridge extending into each side of the tentorium. The vertical diameter of the cerebellum and medulla oblongata equals that of the cerebrum, and is two inches three lines: the transverse diameter of the cerebellum was about three inches nine lines; its antero-posterior extent about one inch and a half. The sculpturing of the internal surface of the cranial cavity bespeaks the high vascularity of the soft parts which it contained, and there are evident indications that the upper and lateral surfaces of the brain had been disposed in a few simple parallel longitudinal convolutions. The two anterior condyloid foramina (*m*) have the same relative position as the single corresponding foramen in the Glossothere, Orycterope, and Armadillos, and the inner surface of the skull slopes outwards from these foramina to the inner margin of the occipital condyle.

Of the bones of the face there remain only portions of the malar, lachrymal, palatine, and maxillaries. The chief peculiarities of the malar bone have been already noticed: the breadth of the base of the descending masseteric processes is two inches two lines; its termination is broken off: the length of the ascending post-orbital process of the malar cannot be determined from the same cause, but it is fortunate that sufficient of this part of the cranium should have been preserved to give this evidence of the affinities of the Scelidothere to the Megathere. The malar bone is continued anteriorly, in a regular curve forwards and upwards, to the lachrymal bone, and completes, with it, the anterior boundary of the orbit: the size of the orbit is relatively smaller than in the Orycterope, and still less than in the Ant-eaters: here, however, we have merely an exemplification of the general law which regulates the relative size of the eye to the body in the mammalia. The malar bone does not extend so far forwards in front of the orbit as in either the Orycterope or Armadillo; in the inclination, however, with which the sides of the face converge forwards from the orbits, the Scelidothere holds an intermediate place between the Armadillos and Orycterope.

The lachrymal bone does not extend so far upon the face in the Scelidothere as in the Orycterope; in which respect the Scelidothere resembles more the Megathere. The foramen for the exit of the infra-orbital nerve has the same situation near the orbit as in the Megathere; its absolute distance from the anterior border of the orbit is only half that in the Orycterope. The foramen is single in the Scelidothere, as in the Orycterope; in the Megathere there are two or three antorbital foramina. The vertical diameter of this foramen is eight lines, the transverse diameter four lines. So much of the outer surface of the superior maxillary bones as has been preserved, is smooth and vertical. Each superior maxillary bone contains the sockets of five teeth, occupying an antero-posterior extent of three inches seven lines, (Pl. XXII and XXIII. fig. 3). The posterior alveolus is situated just behind the transverse line, extending across the anterior boundary of the orbits; the remaining sockets of the molar series extend forwards three inches in front of the orbits. In the Megatherium, the roots of the five superior molars are all situated behind the anterior boundary of the orbit: in the Orycteropus, on the contrary, the grinders are all placed in advance of the orbit; so that the Scelidothere resembles that species more than the Megathere in the relative location of the teeth. The palatal interspace between the roots of the last molar tooth of each series is eleven lines; the palate gradually though slightly widens, as it advances forwards: the posterior margin of the palate is terminated by an acute-angled notch. In the breadth of the bony palate the Scelidothere is intermediate between the Megathere and Orycterope.

The anterior of the upper molars is represented at fig. 3, 4, and 5, Pl. XXI., and at 1, fig. 3, Pl. XXIII.; it corresponds closely in form and size with the opposite molar below; the base of the triangle given by its transverse section is turned inwards and obliquely forwards.

The second molar of the upper jaw, also presents in transverse section a triangular form, with the angles rounded off; but the inner side of the tooth is traversed by a longitudinal groove. The largest diameter of the transverse section, which is placed obliquely as regards the axis of the skull, measures ten lines and a half; the opposite diameter of the tooth is six lines.

The third and fourth molars present the same form and size, and relative position as the second.

The fifth molar is the smallest of the series; its transverse section gives an inequilateral triangle, with the corners rounded off; the broadest side is turned outwards, and is slightly concave; the antero-posterior diameter of this tooth is seven lines; the transverse four lines. The length of the teeth in the upper jaw is about two inches and a half.

It is almost superfluous to observe that the teeth of the Scelidothere, as in other *Bruta*, are without fangs, and have their inserted base excavated by large conical cavities, for the lodgment of a persistent pulp. The tooth is composed of a small central body of coarse ivory or 'dentine,' traversed by medullary canals, which at the periphery of the coarse dentine anastomose by loops, from the convexity of which the calcigerous tubes are given off which form the fine dentine: the layer of this substance, which immediately surrounds the coarse dentine, is about one line and a half in thickness, and the whole is invested with a very thin coating of cement. The teeth of the Scelidothere thus present a more resisting structure than do those of the Mylodon; having a larger proportion of the dense ivory composed of the minute calcigerous tubes, and a much smaller proportion of the softer external cæmentum; in this respect the Scelidothere recedes farther from Megathere, and approaches nearer the Armadillos than does the Mylodon.

The lower jaw resembles, in the general form of the posterior moiety which is here preserved, that of the Sloth and Mylodon more than that of any other Edentate species. Its deep posterior angle is produced backwards, and a broad coronoid process rises and nearly fills the zygomatic space; the condyle is flat, as the glenoid surface has already indicated; its transverse diameter is an inch and eight lines; its antero-posterior diameter seven lines: it is principally extended inwards beyond the vertical line of the ascending ramus. The lower contour of the jaw describes an undulating line; which, commencing from the posterior angle, is at first gently convex, then slightly concave, then again convex, below the alveoli of the teeth, where it is rounded and expanded, as in the Orycterope. The fractured condition of the right ramus of this part fortunately exposed the roots of the four grinding teeth, which constitute the dental series on each side of the lower jaw. The length of the jaw occupied by these four alveoli is three inches ten lines, which exceeds a little that of the opposed five grinders above; the ramus of the jaw gradually diminishes in all its dimensions anterior to the molar teeth; the dental canal passes in a gentle curve below, and on the inner side of the alveoli, whence it gradually inclines to the outer wall of the jaw.

The whole ascending ramus of the jaw consists of a very thin plate of bone; it is slightly concave on the inner side, and the inferior margin of the produced angle inclines inwards, as in the Mylodon and Sloth; it is impressed on the outer side with two shallow depressions, and two parallel ridges, both following the gentle curvature of the part. There is a foramen on the outer side of the ramus at the anterior part of the base of the coronoid process corresponding with that in the lower jaw of the Mylodon, but the longitudinal channel which runs along the outer side of the alveolar processes is wanting, and the expansion at the base of those processes is more sudden and relatively greater; the general correspondence, however, between these lower jaws is such as would lead to the idea that they belonged to animals of the same genus, were it not that the teeth present modifications of form in the Scelidothere, as distinct from those of the Mylodon, as are any of the minor dental differences on which genera or sub-genera of existing Mammalia are founded in the present state of

Zoological Classification.

To make this distinction more readily intelligible, I have given a view of the transverse section of the teeth in the right ramus of the lower jaw (fig. 4, Pl. XXIII.), corresponding with that of the *Mylodon Darwinii*, (Pl. XVII., fig. 5). In the present sub-genus the antero-posterior extent of the four alveoli of the lower jaw nearly equals four inches, and is relatively greater than in the Mylodon, although the teeth are placed closer together; this is owing to their greater relative size. The first molar tooth presents the simplest form; its transverse section is a compressed inequilateral triangle with the angles rounded off; the longest diameter of this section which is parallel with the inner alveolar border is eleven lines, the transverse diameter almost six lines; the base or broadest side of the triangle is turned inwards, and is slightly concave; the two smaller sides are also slightly concave.

The second molar is placed more obliquely in the jaw; the long axis of its transverse section intersects at an acute angle that of the jaw itself; the transverse section presents a compressed or oblong form, with the larger end next the outer side, and the smaller end next the inner side of the jaw; this end is simply rounded, but the outer end presents a sinuosity, corresponding to a broad groove which traverses the whole length of the outer side of the tooth; the anterior, which corresponds to the internal side or base of the transverse section of the preceding molar, is slightly concave.

The third molar has nearly the same form and relative position as the preceding; the long diameter of the transverse section is, in both, ten lines and a half; the principal transverse diameter is, in the second molar five lines, in the third nearly six; the difference of form observable in these as compared with the two middle grinders of the Mylodon is well marked; in the latter these teeth are impressed with a longitudinal groove on their inner sides; in the Scelidothere they have a similar impression along their outer but not along the inner side.

In the last molar the resemblance is much closer, and the

modification of form by which it differs from the preceding ones is of the same kind; the transverse section gives an irregular oblong figure with its axis nearly parallel with that of the jaw, and constricted at the middle by sinuosities produced by two wide channels which traverse longitudinally, one the outer, the other the inner side of the tooth; the latter groove is much wider and shallower in the Scelidothere than in the Mylodon. The two lobes produced by these grooves are more equal in Scelidothere; the anterior one is concave on its anterior surface instead of convex as in the Mylodon; the posterior one is more compressed; the longitudinal or antero-posterior diameter of the transverse section of this tooth is one inch five lines; the greatest transverse diameter is nine lines; the diameter of the isthmus joining the lobes is three lines and a half; the entire length of this tooth is three inches three lines.*

VERTEBRAL COLUMN.

Of this part of the skeleton of the Scelidothere, Mr. Darwin's specimen includes, as is represented in Plate XX., the cervical, part of the dorsal, and the sacral series of vertebræ in a more or less perfect condition.

The cervical vertebræ present the ordinary mammalian number, seven, and are free, or so articulated as to have permitted reciprocal movement upon each other. Their transverse processes are perforated as usual for the vertebral arteries. These processes in the atlas are remarkable for their great breadth, length, and thickness; and indicate the muscular forces which must have worked the head upon the spine to have been very powerful. The axis is provided with a robust 'processus dentatus,' having a base equal in breadth to the body of the axis itself; and a smooth

---

* It requires little stretch of imagination to conceive that this more complex posterior tooth (Pl. XXIII, fig. 4, 4) in the lower jaw is the representative of the two smaller posterior teeth (ib. fig. 3, 4, and 5) of the upper jaw conjoined.

articular convexity on the side of the apex on which the ring of the atlas rotated. The line of union between the axis and its characteristic process, which here resembles the body of an abortive vertebra, is very distinct. The transverse processes of the vertebra dentata are comparatively feeble, but this condition is amply compensated for by the great development of the spinous process. (Pl. XXIV. fig. 1.) This process is bent backwards at nearly a right angle, overlaps with its reflected extremity the spine of the third cervical vertebra, and rests by its base, on the under part of which are the posterior articular surfaces, upon the broad and strong anterior oblique processes of the third vertebra.

The third, fourth, fifth, and sixth cervical vertebræ have moderately developed and pointed spinous processes: their transverse processes are broad, and extend obliquely backwards, and slightly overlap each other. On the under part of the transverse process of the sixth cervical vertebra there is the fractured base of what I conjecture to have been an expanded aliform plate, analogous to that observable in the corresponding vertebra of the Orycterope. The seventh cervical vertebra has part of the articular depression for the head of the first rib upon each side of its body: the transverse process is feebly developed, but the spine is double the height and size of those of the preceding vertebræ.

The spinous process of the first dorsal vertebra in like manner rises to twice the height of the preceding spine of the seventh cervical, and preserves an equal antero-posterior diameter from its base to its summit, which is thick and slightly bent backwards: four or five succeeding dorsal vertebræ give evidence of having been surmounted by spines of equal height and strength. The transverse processes of these dorsal vertebræ present bold concavities on their inferior part for the reception of the tubercles of the ribs, and they gradually ascend upon the base of the spines as the vertebræ are placed further back, so as to increase the expansiveness of the chest. The state of the fossil did not afford further information as to the condition of this part of the vertebral column, but the parts which have been preserved are precisely those from which the most interesting inferences as to the affinities and habits of the extinct quadruped can be deduced.

Whether the Megatherium be most nearly allied to the tribes of the Sloth or Armadillo has been a question under recent discussion, and, as a corollary of this problem, whether its habits were those of a scansorial or of a fossorial quadruped. For, strange as it may appear at first sight, there have not been wanting arguments, and those urged by an anatomist to whom we owe much novel and interesting information respecting the extinct Edentata, in support of the belief that the Megatherium, gigantic and ponderous as must have been its frame, actually climbed trees like a Sloth, and had claws and feet organised for prehensile actions, and not in accordance with that type by which they are usually adapted for digging up the soil.*

Now, in whatever degree the Megatherium may be involved in this question, the smaller Megatherioid species at present under consideration must be at least equally implicated in it. In the adaptation of the frame of a mammiferous quadruped for especial and peculiar actions and modes of life, such as for climbing and living in trees, or for burrowing and seeking concealment in the earth, not only the immediate instruments, as the feet, are modified, but the whole of the osseous and muscular fabric is more or less impressed with corresponding adaptations, whilst at the same time these special adjustments are invariably subordinated to the type of organization which characterizes the group.

The type of the order *Bruta* or *Edentata* is well-marked; one or more claws of unusual length and strength, characterize the forefeet and sometimes the hind-feet in every genus, and the term 'Macronykia' would more aptly designate them than the term which Cuvier substituted for the good old Linnæan appellation.

---

* Lund, Videnskabernes Selskabs, Natur: og Mathem. Afhandlinger, Kiöbenhavn, vol. viii.

The uniform absence of true roots to the teeth, where these are present, is another general character; the skeleton exhibits many well-marked peculiarities common to the whole order; while at the same time it is modified in various modes and degrees in accordance with the peculiar habits and exigencies of the species.

One of the regions of the skeleton which manifests adaptive modifications of this kind in the most remarkable degree is the cervical division of the vertebral column. In one edentate species it is lengthened out by two additional vertebræ more than in any other mammal; in another it is reduced by anchylosis to as great an extent below the regular number of moveable pieces: and these, the two most opposite conditions of the cervical vertebræ which are to be met with in the mammiferous class are related to equally diverse and opposite habits of life.

With respect to the *Ai*, or three-toed Sloth, "an animal, great part of whose life, when not engaged in eating, is spent in sleeping on trees,—an easy attitude for repose is most essential to its comfortable existence; and accordingly we find, that the auxiliary vertebræ at the base of the neck contribute to produce that flexibility of this organ which allows the head of the animal to incline forwards and rest upon its bosom." Dr. Buckland, from whose Paper on the "Adaptation of the Structure of the Sloths to their peculiar Mode of Life,"* the preceding judicious physiological remark is quoted, adduces the authority of Mr. Burchell in proof that the Sloth can in a remarkable manner and with great facility twist its head quite round, and look in the face of a person standing directly behind it, while at the same time the body and limbs remain unmoved. A single glance at the length and slenderness of the cervical region of the spine, and of the feeble condition of the transverse and spinous processes in the vertebræ composing that part of the skeleton of the Sloth, is enough to show its adaptation to increase the rotatory motion

---

* Linn. Trans, vol. xvii. (1833) p. 17.

and flexibility of the neck.

In describing the skeleton of a species of Armadillo (*Dasypus* 6-*cinctus*, Linn.)† I was led in like manner to point out the subserviency of the peculiarities of the cervical vertebræ to the habits and mode of life of that animal; observing that the "anchylosis of the cervical vertebræ obtains in the *Cetacea*, as well as in the genus *Dasypus*, and that as in the aquatic order this firm connexion of the cervical vertebræ assists materially in enabling the head to overcome the resistance of the dense fluid through which they perpetually move, so in the Armadillos a like advantage may be derived from this structure during the act of displacing the denser material in which they excavate their retreats."‡

Having in view these well-marked examples of the subserviency of the structure of the bones of the neck to the habits of existing species of the order *Bruta*, I proceeded to investigate the structure of the corresponding part of the skeleton in the *Scelidotherium*, hoping thereby to gain a new and useful element in the determination of the problem at present under discussion, as to the affinities and habits of the extinct Megatherioid quadrupeds.

The fossil, in its original state, yielded a view of so much of the anterior part of the bodies of the cervical vertebræ as proved that they were neither so numerous as in the Sloth, nor anchylosed together as in the Armadillos: after a long and careful chiselling at the hard matrix in which they were imbedded, the transverse and spinous processes were exposed to view, as they are represented in Plates XX. and XXIV. The description of these

---

† Zool. Proceedings, 1832, p. 134.
‡ The anterior prolongation of the sternum in front of the neck and the corresponding anterior position of the clavicles and scapulæ occasions a transference of such a proportion of the moving powers of the head from the cervical vertebræ to these bones in the mole, as renders any modifications of these vertebræ, like those in the Armadillo, uncalled for.

processes has already been given.

On comparing the cervical vertebræ of the Scelidotherium with those of the existing *Bruta*, the closest resemblance to them was found in the skeleton of the Orycterope. Now this quadruped, though not so rapid a burrower, or so strictly a subterranean species as the Armadillos, participates, nevertheless, to a certain extent, in their fossorial habits, and is closely allied to them in general structure: it differs from them, indeed, mainly in a modification of the dental system, in the absence of dermal armour, and of anchylosis of the cervical vertebræ. But the advantages which, as a burrower, it would have derived from the latter structure, are compensated for by the shortness of the cervical vertebræ, and by the great development and imbricated or interlocking co-adaptation of the transverse and anterior spinous processes of the cervical vertebræ. The analogous quadruped in the South American Continent—the great ant-eater (*myrmecophaga jubata*) which uses its powerful compressed fossorial claws for breaking through the hard walls of the habitations of its insect prey, but which does not excavate a subterraneous retreat for itself, presents the cervical vertebræ of a more elongated form, and without that development of the spinous and transverse processes which tend to fix the neck and increase the size of the muscles which move the head: and, if we could conceive that its fore-feet were employed to scratch up vegetable roots, instead of disinterring termites, there would be no reason to expect any modification of the cervical vertebræ as a direct consequence of such a difference in the application of its fossorial extremities: when, therefore, we find that the cervical vertebræ do actually differ in two myrmecophagous species, to the extent observable in the Cape and South American ant-eaters, we arrive legitimately at the conclusion that such difference relates to fossorial habits of the one species, in which habits the other does not participate.

Now, therefore, if this conclusion be just in regard to the Orycterope, it must bear with more force upon the question of the habits of the Scelidotherium as the mechanism for strengthening the connection of cervical vertebræ, and for augmenting the surface of attachment of the muscles which worked the head and neck, is more strongly wrought out in that extinct species.

The great size and strength of the spinous process of the dentata, and the mode in which it is interlocked with the spinous and oblique processes of the third cervical, together with the imbricated disposition of the transverse processes of this and the succeeding vertebræ, and the remarkable height of the dorsal spines, all combine to indicate in a very striking manner, if not to demonstrate, that the conical head of the present species, which is comparatively small and slender, and

for its own mere support requiring therefore no such mechanism, was used in aid of the fossorial actions of the extremities.

As the cervical vertebræ of the Megatherium have their processes comparatively weaker than in the Scelidotherium, and the anterior dorsal spines are relatively shorter, it may be concluded, that whatever were the extent or nature of the fossorial labours of the enormous claws with which it was provided, the head did not co-operate with the digging implements in their especial task in the same degree as in the Scelidothere and Orycterope. At the same time there is no modification of the cervical region of the spine of the Megathere corresponding with those which we have seen to be subservient to the arboreal habits of the sloth, a remark which will not be deemed superfluous by those who have perused the acute observations and arguments adduced by M. Lund in favour of the scansorial character of the extremities of the Megatherium and Megalonyx.

The fragments of the dorsal vertebræ and ribs of the Scelidotherium, which are figured in Plate XX, offer no modifications which need detain our attention; they closely conform, excepting in the greater relative height of the anterior dorsal spines, already noticed, with the Megatherioid type. The sacrum manifests in its vast expanse, the great development of the posterior transverse processes to join the ischium, the capacious medullary cavity, and wide nervous foramina, a like conformity with the Megatherium, and a corresponding harmony with the disproportionate bulk of the hind legs.

BONES OF THE EXTREMITIES.

The Scapula in its double spine, the osseous arch formed by the confluence of the acromion with the coracoid process, and the substitution of a distinct foramen for the suprascapular notch, agrees with that of the Megatherium: but the span of the acromial arch is relatively wider, and the surface for the articulation of the clavicle is better marked. This articular surface, which is distinctly shewn upon the acromion of both the scapulæ in Pl. XX. is the more interesting, as being the only evidence of the clavicle of the Scelidothere which we at present possess; but it is enough to prove that this quadruped enjoyed all the advantages in the actions of the fore-extremity, which arise out of the additional fixation of the shoulder-joint afforded by the clavicle—a bone which the extinct Megatherioids are the largest of the mammiferous class to possess in a completely developed state. The form, position, and aspect of the glenoid cavity for the humerus closely correspond with the condition of the same part in the Megatherium. The limits of the acromial and coronoid portions of the arch were still defineable in the present skeleton, which indicates the nonage of the individual in the unanchylosed condition of most of the epiphyses.

In regard to the presence of a clavicle in the Megalonyx M. Lund has deduced certain conclusions, which, if well founded, would be equally applicable to the present allied species, and to the great Megatherium. I am induced, therefore, to offer a few physiological observations on that bone, which appear to me to lead to a more correct interpretation of its uses and relations in the great mammiferous animals now under consideration.

When the anterior extremities in mammalia are used simply for the purpose of progressive motion on dry land, as in the Pachyderms and Ruminants, or in water, as in the Cetaceans, there is no clavicle; this bone is introduced between the sternum and acromion, in order to give firmness and fixity to the shoulder-joint when the fore-leg is to discharge some other office than that of locomotion. In these cases, however, the clavicle exists in various degrees of development, and even its rudiment may be dispensed with in some of the actions which require a considerable extent of lateral or outward motion, and of freedom of rotation of the fore-limb. When, therefore, we find the clavicle fully developed in the skeleton of an extinct mammiferous animal, and so placed as to give the humeral articulation all the benefit of this additional mechanism, we may confidently expect that it will afford an insight into the habits and mode of life of such extinct species. M. Lund* has argued from the clavicle of the Megalonyx, that it climbed like a Sloth. "Animals," says Sir C. Bell,† "which fly or dig, or climb, as Bats, Moles, Porcupines, Squirrels, Ant-eaters, Armadilloes, and Sloths, have this bone; for in them, a lateral or outward motion is required." But in regard to the present problem, we have to enquire whether the clavicle manifests any modifications of form, of strength, or development in relation to the special differences of these several actions, with which its presence is asserted to be associated?

In mammals which fly, the clavicle is always complete: the rabbit, the fox, and the badger are instances of burrowing animals in which the clavicle is absent or rudimental. The presence of a perfect clavicle is not more constant in climbing quadrupeds. The Ai, for example, has an incomplete clavicle, which is attached to the acromion process, and terminates in a point about one-fourth of the distance between the acromion and the

---

*   Loc. cit.
†   Bridgewater Treatise, p. 46.

top of the sternum, to which the clavicular style is attached by a long slender ligament: the advantage, therefore, which a perfect clavicle affords in the fixation of the shoulder-joint, is lost to this climber *par excellence*. Again, the Bears, which are the bulkiest quadrupeds that are gifted with the faculty of climbing, and this in so perfect a degree that the Sun-bears of the Eastern Tropics may be termed arboreal animals,—these scansorial quadrupeds are destitute of even the smallest rudiment of a clavicle, as I have ascertained by repeated careful dissection.

Since, therefore, a clavicle in any degree of development is not essential to a climbing quadruped, we must seek for some other relation and use of that remarkably strong, and perfect bone, as it exists in the Megathere, Megalonyx, and Scelidothere. The absence of 'dentes primores' or of anterior or incisive teeth in these quadrupeds at once sets aside any idea of its connection with an action of the fore extremities, very common in the mammals which possess clavicles, viz., that of carrying the food to the mouth, and holding it there to be gnawed by the teeth. Flying is of course out of the question, although our surprise would hardly be less at seeing a beast as bulky as an elephant climbing a tree, than it would be to witness it moving through the air. If now we restrict our comparison to the relations of the clavicle in that order of Mammalia to which the extinct species in question belonged, we shall see that it is most constant, strongest, and most complete in those species which make most use of their strong and long claws in displacing the earth, as the Armadilloes and Orycteropus: and, as the clavicle is incomplete in one climbing Edental, we are naturally led to conclude that its perfect development in an extinct species must have been associated with uses and relations analogous to those with which it coexists in other genera of the same order. Thus it will be seen, that, in rejecting the conclusion drawn by M. Lund from the presence of a clavicle, I concur in the opinion expressed by Dr.

Buckland[*] that the Megatherium—and with it the Megalonyx and Scelidotherium—had the shoulder-joint strengthened by the clavicle, in reference to the office of the fore-arm, as an instrument to be employed in digging roots out of the ground. Not, however, that these gigantic quadrupeds fed on roots, but rather, as the structure of the teeth would show, on the foliage of the trees uprooted by the agency of this powerful mechanism of the fore-legs, and of the otherwise unintelligible colossal strength of the haunches, hind-legs, and tail.

The humerus presents a large convex oval head, on each side of which is a tuberosity for the implantation of the supra- and sub-scapular muscles: these tuberosities do not rise above the articular convexity, so as to restrict the movements of the shoulder-joint, as in the Horse and Ruminants, but exhibit a structure and disposition conformable to those which characterize the proximal extremity of the humerus in other mammalia which enjoy rotatory movements of the upper or fore-limb. The tuberosities are, however, relatively more developed, and give greater breadth to the proximal end of the humerus in the Scelidothere than in the Megathere. The distal end of the humerus, although mutilated, clearly indicates that it had the same characteristic breadth of the external and internal condyles, as in the Megatherium. In fig. 1. Pl. XXV. which gives a front view of the left humerus, the broad internal condyle, with its extremity broken off, is seen projecting to the left hand; both in this figure and in fig. 2. in which the internal side of the humerus is turned towards the observer, the wide groove, with its two osseous boundaries, is shewn, which plainly indicates that the left condyle was perforated for the direct passage of the artery or median nerve, or of both, to the fore-arm. The groove for the musculo-spiral nerve on the outer side of the humerus is over-arched at its upper part by a strong obtuse process; which is comparatively less developed in the Megatherium. The trochlear

---

[*] Bridgewater Treatise, p. 152.

or inferior articular surface of the humerus presents, as in the Megatherium, two well marked convexities, with an intervening concavity: this indication of the rotatory power of the fore-leg is confirmed by the form of the head of the radius.

In Pl. XXV. fig. 4. a view is given of this articular surface: it presents the form of a subcircular gentle concavity, which plays upon the outer convexity of the humeral articular surface: immediately below the upper concavity the radius presents a lateral smooth convex surface, which rotates upon a small concavity on the ulna, analogous to the 'lesser semilunar,' in human anatomy, in which the mechanism for rotation, so far as the upper joint of the radius is concerned, is not more elaborately wrought out than in the present extinct edentate quadruped. The radius expands as it proceeds to the elbow-joint, where it attains a breadth indicative of the great power and size of the unguiculate paw, of which it may be called the stem, and to the movements of which it served as the pivot.

All the bones of the fore-limb just described—the scapula, the humerus, and the radius,—indicate by the bold features and projections of the muscular ridges and tubercles the prodigious force which was concentrated upon the actions of the fore-paw, and the ulna, in its broad and high olecranon (of which a side-view is given in fig. 2. Pl. XXV.) gives corresponding evidence. The great semilunar concavity is traversed by a sub-median smooth ridge, which plays upon the interspace of the two humeral convexities. The body of the bone is subcompressed, straight, and diminishes in size as it approaches the carpal joint: the immediate articulating surfaces are wanting in both the radius and ulna, the epiphysial distal extremities having become detached from their respective diaphyses.

Of the terminal segment of the locomotive extremities, the only evidence among the remains of the skeleton of the Scelidothere is the ungueal phalanx figured at Pl. XXVII. 3, 4, and 5; but as it is uncertain whether it belong to the fore or hind-foot, it will be described after the other bones of the extremities have been noticed.

Of these bones the femur is the most remarkable, both for its great proportional size, and its extreme breadth, as compared with its length or thick-ness: but in all these circumstances the affinity of the Scelidothere with the Megathere is prominently brought into view. There is no other known quadruped with which the Scelidothere so closely corresponds in this respect. In proceeding, however, to compare together the thigh-bones of these two extinct quadrupeds, several differences present themselves, which are worthy of notice: of these the first is the presence in the Scelidothere of a depression for a 'ligamentum teres' on the back part of the head of the femur, near its junction with the neck of the bone: this is shewn in the posterior view of the femur given in Pl. XX. The head itself forms a pretty regular hemisphere: the great trochanter does not rise so high as in the Megatherium, but, relatively, it emulates it in breadth: the small trochanter is proportionally more developed: the external contour of the shaft of the femur is straighter in the Scelidothere than in the Megathere, and the shaft itself is less bowed forwards at that part. The articular condyles occupy a relatively smaller space upon the distal extremity of the femur in the Scelidothere, and they differ more strikingly from those of the Megathere, in being continued one into the other: the rotular surface, for example, which is shewn in fig. 5. Pl. XXV. is formed by both condyles, while in the Megatherium it is a continuation exclusively of the external articular surface.

The patella, which works upon the above-mentioned surface, is a thick strong ovate bone, with the smaller end downwards: rough and convex externally, smooth on the internal surface, which is concave in the vertical and convex in the transverse directions.

Of the bones of the leg only the proximal end of the tibia is preserved; but this is valuable, as shewing another well-marked difference between the Scelidothere and Megathere; for whereas in the latter the fibula is anchylosed with the tibia, this bone, in the Scelidothere, presents a smooth flat oval articular surface, which is shewn in fig. 2. Pl. XXVII. below the outer part of the head of the bone; from the size and appearance of which, I infer, that the fibula would not have become confluent with the tibia, even in the mature and full-grown animal.

The relative length of the fore and hind extremities cannot be precisely determined from the present imperfect skeleton of the Scelidothere; but there is good evidence for believing, that the fore extremity was the shortest. The humerus is shorter than the femur by one-ninth part of the latter bone; and the radius, which wants only the distal epiphysis, must have been shorter than the humerus. Now the relative development of the fore and hind legs is one of the points to be taken into consideration in an attempt to determine the habits and nature of an extinct mammal.

In climbing animals the prehensile power is more essential to the hinder than to the fore parts or extremities. In the leech the principal sucker is in the tail; and higher organized climbers, in like manner, depend mainly on their posterior claspers in descending trees, and hold on by means of them whilst selecting the place for the next application of those at the fore part of the body, whether their place be supplied by the beak, as in the Maccaws, or the fore-feet or hands in the Mammalia.

But, although we perceive the hinder limbs to be the last to lose the advantageous structure of the hand in the Quadrumanous species, and notwithstanding that the tail is for this purpose sometimes specially organized to serve as a prehensile instrument, yet we find that the power of grasping the branches of trees by either legs or tail is never maintained at the expense of undue bulk and weight of those organs. On the contrary, as the fore-limbs are the main instruments in the active exertions of climbing, so they are the strongest as well as the longest in all the best climbers, and the weight of the body which they have to drag along is diminished by dwarfish proportions of the hinder limbs, as in the Orangs and the Sloths.

Can those huge quadrupeds have been destined to climb that had the pelvis and hinder extremities more ponderous and bulky in proportion to the fore-parts of the body than in any other known existing or extinct vertebrate animals?

M. Lund argues for the scansorial character of the Megalonyx, because its anterior extremities are longer than the posterior ones; but if they somewhat exceed the hind-legs in length, how vastly inferior are they in respect of their breadth and thickness. The prehensile faculty of the hinder limbs of the best climbers, as the Sloths, Orangs, and Chameleons is by no means dependent on the superior mass of muscle and bone which enters into their conformation, but is associated with the very reverse conditions.

It is impossible to survey the discrepancy of size between the femur and the humerus of the Scelidothere, as exhibited in Pl. XX., without a conviction that it relates to other habits than those of climbing trees. The expanse of the sacrum, the evidence of the muscular masses employed in working the hind legs and tail, which is afforded by the capacity of the cavity lodging the part of the spinal marrow from which the nerves of those muscles were derived, both indicate the actions of the hind-legs and tail to have been more powerful and energetic than would be required for mere prehension: and the association of hinder extremities so remarkable for their bulk, with a long and powerful tail, forbids my yielding assent to the speculation set forth by M. Lund, as to the prehensile character of the tail of the Megalonyx.

Astragalus.—In the examination of this characteristic bone I have kept in view the question of the habits of the Megatherioid quadrupeds in general, and the especial affinities of the Scelidotherium, in illustration of which I shall notice at the same time the peculiarities of the astragalus of the Sloth, Megatherium and Armadillo.*

The upper articular surface of the astragalus of the Scelidotherium (Pl. XXVI. fig. 4.), presents, in its transverse contour, two convex pulleys, *a* and *b*, and an intermediate

---

* *Dasypus 6-cinctus*, L., is the species of which I have the astragalus separate, so as to be able to follow out the comparison.

concavity, forming one continuous articular surface. The external or fibular trochlea (*a*) is strictly speaking convex only at its posterior part, the upper surface gradually narrowing to a ridge, as it advances forwards, from which, the inner and outer parts slope away at an angle of 35°.

The tibial* convexity (*b*) is more regular and less elevated, it has only half the antero-posterior extent of the outer pulley; its marginal contour forms an obtuse angle at the inner side.

In the Megatherium the upper articular surface of the astragalus is also divided into two trochleæ, of which the one on the fibular side (fig. 3, a), is of much greater relative size and extent than the tibial one (b), and is raised nearly four inches above the level of the latter, although in the oblique position in which the bone is naturally placed in the skeleton, the highest part of each convexity is on the same level The fibular trochlea differs also from that in the Scelidothere in being regularly convex in the transverse as well as the antero-posterior direction. The tibial convexity resembles that in the Scelidothere, save in its smaller relative size; its internal margin likewise forms an angular projection below the internal malleolus.

The upper surface of the astragalus of the Mylodon, or Megalonyx(?) (Pl. XXVIII. fig. 5.),† differs from that in the Megatherium in having a narrower fibular trochlear ridge.

The astragalus of the Ai (*Bradypus tridactylus*) differs widely from that of either the Megathere, Mylodon (?) or Scelidothere in having a conical cavity on the upper surface, in place of the fibular convexity, in which concavity the distal end of the fibula

---

* In distinguishing these trochleæ as fibular and tibial, it is to be understood that the terms relate only to aspects corresponding to the position of those bones, and not that the fibula is articulated to the whole of the trochlea so called: it probably rested only upon the outer facet in the Scelidothere.

† This astragalus was found at Santa Fé, in Entre Rios, associated with the remains of the Mastodon and Toxodon; but from its size and form I entertain little doubt that it belonged to a Megatherioid quadruped as large as the Mylodon or Megalonyx. The brief allusion to the astragalus of the Megalonyx in M. Lund's Memoir does not afford the means of determining with certainty this point.

rotates like a pivot. This mechanism is closely related to the scansorial uses of the inwardly inflected foot of the Sloth.

If the astragalus of an Armadillo[‡] were placed side by side with that of the Megathere, it would be very difficult to determine the analogous parts, especially of the upper surface, unless guided by the intermediate structure presented by the Scelidothere. The upper surface of this bone, in the Armadillo, is, however, divided into two transversely convex trochleæ, separated by a much wider transversely concave surface. The fibular trochlea resembles that of the Scelidothere in having its upper and outer facets sloping away at an acute angle, but without meeting at a ridge anteriorly; this surface is not more raised above the tibial trochlea than in the Scelidothere.

The inner trochlea differs from that of the Scelidothere in having a greater relative antero-posterior extent, and in forming, in place of an uniform convex surface, a trochlea similar in structure to that on the outer side. The extent of rough surface on the upper part of the astragalus intervening between the articular surface for the bones of the leg, and that for the scaphoides is extremely small in the Megathere and Mylodon (?); it is relatively greater in the Scelidothere; it is still more extensive in the Armadillo; but is the longest in the Sloth. The anterior extremity of the astragalus which is entirely occupied by the scaphoid articular surface is very peculiar in the Scelidothere (Pl. XXVI. fig. 2.): it presents one convex and two concave facets, which, however, form part of one continuous articular surface: the convex facet forms the internal part of the surface, and presents a rhomboidal form with the long axis vertical. The concave facets (*c* and *d*) are extended transversely and placed one above the other; they are slightly concave in the transverse, and nearly flat in the vertical directions.

In the Megatherium (fig. 1.) the scaphoid surface of the

[‡] See the figures of this bone, given by Cuvier in Pl. x. and xi. Ossemens Fossiles, vol. v. part i.

astragalus is divided only into one concave and one convex portion, both continuous with each other: the concave facet (*c*) corresponds with the upper concavity in the Scelidothere, but is a pretty uniform subcircular depression, fourteen lines in depth: the convex facet, *d*, is continued across the whole breadth of the under part of the scaphoid surface and corresponds with both the inner convex, and lower concave surfaces of the scaphoid articulation in the Scelidothere.

In the Mylodon (?) (Pl. XXVIII. fig. 3.), the articular facet, corresponding with that marked (*c*) in the astragali of the Megathere and Scelidothere, is simply flattened, instead of being concave; the rest of the scaphoid surface corresponds with that in the Megatherium.

In the Armadillo the scaphoid articular surface is undivided and wholly convex: in this part of the astragalus, therefore, we find the Scelidothere deviating from the Armadillo further than does the Megathere; while the Mylodon or Megalonyx (?) most resembles the Armadillo in the configuration of this part of the astragalus.

If we compare the outer surfaces of the astragalus in these quadrupeds, we shall find, however, that the Scelidothere and Armadillo closely agree: the outer facet of the fibular trochleæ, above described, is continued in the Scelidothere (Pl. XXVIII. fig. 2.), upon the fibular side of the astragalus reaching nearly half-way down the posterior part, and down nearly the whole of its anterior.

In the Armadillo, it extends over the whole of the anterior part of the outer side of the astragalus. In both animals the lower boundary of this articular surface describes a strong sigmoid curve.

In the Megatherium (Pl. XXVIII. fig. 1), the corresponding surface for the fibular malleolus on the outer side of the astragalus is formed by a comparatively very small semicircular flattened facet, which by its roughness indicates that the end of the fibula was attached to it by ligamentous substance, and that the synovial bag was not continued upon that surface as in the Scelidothere and Armadillo.

In the Mylodon (?) (Pl. XXVIII. fig. 4), even this rough facet is wanting and the fibular trochlea is bounded by the angle which divides the upper from the outer surface of the astragalus.

Turning now our attention to the under surface of the astragalus, we observe that it presents in the Scelidothere (Pl. XXVI. fig. 6), an irregular quadrate form, having the outer side occupied by an elongated sub-ovate articular facet, e, for the calcaneum, bounded externally by a sharp edge, with its long axis and its greatest concavity in the antero-posterior direction, and slightly convex from side to side: a second calcaneal articular surface (f) is situated at the inner and anterior angle; it is oblong and nearly flat; is continuous with the inferior concave facet of the scaphoid articulation, but is divided from the convex facet by a groove: the two calcaneal articulations are separated by a deep and rough depression, traversing the under surface of the astragalus diagonally, and increasing in breadth towards the posterior and internal angle. The inner side of the astragalus presents a convex protuberance.

The correspondence between the astragalus of the Scelidothere and Megathere is best seen at the under surface of the bone: in both the two calcaneal articulations are separated by the diagonal depression, and the internal and anterior surface is continuous with the scaphoid articulation. In the Megathere, however, in consequence of the absence of the inferior concavity which characterizes the Scelidothere, the anterior calcaneal facet (f) appears as a more direct backward continuation of the scaphoidal surface; but they are divided by a more marked angle than is represented in the figure (fig. 5, Pl. XXVI.). The posterior and outer calcaneal surface in the Megathere (e) is broader in proportion to its length, continued further upwards upon the outward surface, is consequently more convex in the transverse direction, and is not bounded externally by so sharp and prominent a ridge as in the Scelidothere. The protuberance from the inner surface of the astragalus is more compressed laterally in the Megathere than in the Scelidothere. The correspondence between the astragali of the Mylodon (?) (Pl. XXVIII. fig. 6) and Megathere in the conformation of the under surface is so close, that the few differences which exist will be sufficiently appreciated by an inspection of the figures.

In the Armadillo the astragalus, in consequence of the greater production of its anterior part, presents more of an angular than a quadrate figure; and the scaphoid articular surface, being proportionally carried forwards, is altogether separated from the anterior calcaneal surface. The posterior and inner calcaneal surface resembles that in the Scelidothere, but is less inclined upwards; and is continuous with the posterior part of the tibial articular surface.

Thus the astragalus in the structure of its two most important articulations, viz. that which receives the superincumbent weight from the leg, and that which transmits it to the heel, presents a closer correspondence in the Scelidothere with that of the Dasypus, than with that of the Megathere or Mylodon.

The ungueal phalanx of the Scelidothere before alluded to, is represented of the natural size in Pl. XXVII. The side-view, fig. 3. shows the position of the articular surface on the proximal end, sloping obliquely towards the under surface, and overtopped by an obtuse protuberance, calculated to impede any upward retraction of the claw: the present joint, in fact, illustrates in every particular the argument by which Cuvier established the true affinities of the allied extinct genus Megalonyx.[*]

The present phalanx is, however, less compressed, and less incurved than those of the Megalonyx, which have been hitherto described; but it more resembles in these proportions one of the smaller, and presumed hinder, ungueal phalanges of the Megatherium. The upper and lateral parts of the bone are rounded, and it gradually tapers to the apex, which is broken off.

---

[*] Ossemens Fossiles, vol. v. part i. p. 163.

The osseous sheath for the claw is developed only at the under part of the bone: it presents the form of a thick flat plate of bone, with the margin very regularly and obliquely bevelled off, and having a vertical process of bone attached lengthwise to the middle of its under surface This process must have served for the insertion of a very powerful flexor tendon. The figures of this bone preclude the necessity of any further verbal description.

M. Lund lays most stress upon the argument founded on the inward inflection of the sole of the foot in the Megalonyx, and appeals with greatest confidence to this structure in support of his hypothesis of the scansorial habits of that extinct Edental.[†]

[†] For the translation of the following passage, and of others alluded to in the present work, from the original Danish Memoir of M. Lund, loc. cit., I am much indebted to the Rev. W. Bilton, M.A. &c. &c.:—

"Thus in every point of comparison we have instituted between the organization of burrowers and climbers; we have seen that the Megalonyx constantly differs from the former and resembles the latter; but the point to which I last alluded (the obliquity of foot), I consider to be quite decisive.

"There is one other point in its organization, which is not quite without weight in reference to our present inquiry,—I mean its unusually powerful tail. Now, it is certainly true that many animals which are not climbers have a powerful tail, as e. g. Armadillos, while the others that climb well, have none, as Sloths and Apes. But when we find a remarkably powerful tail attached to an animal that according to all probability was a climber, we are led to infer that this organ must have served for that purpose: in other words, that the Megalonyx was furnished with a prehensile tail.

"How far the Megatherium is to be considered in the same light as the Megalonyx cannot be decided without an accurate and scientific examination of its skeleton at Madrid. Pander and D'Alton do not mention any distortion of the hind-foot, neither does their figure exhibit any. It is nevertheless quite possible that such may exist, but that it is disguised by the faulty manner in which the skeleton is put up. It strikes me as little probable that two animals which agree so well in the principal particulars of their organization should differ so much in one of the most important. The Megatherium has been proved by later discoveries to possess the same powerful tail as the Megalonyx, and as it corresponds also with the latter entirely in the conformation of its extremities, the same difficulties present themselves against the supposition of its having been a burrower. But if

It is quite true that the Quadrumana derive advantage from this position of the foot in climbing trees, and that it is carried to excess in the Sloths, which can only apply the outer edge of the foot to the ground. But we may ask, was the inversion of the sole of the foot actually carried to such an extent in the *Megalonyx*? And, admitting its existence in an inferior degree, is it then conclusive as to the scansorial habits of that species?

M. Lund expressly states that it is produced by a different structure and arrangement of the tarsal bones, from that which exists in the Sloth, but he does not specify the nature of this difference.

If the astragalus, which I have referred with doubt to the *Megalonyx*, do not actually belong to that genus, it is evidently part of a very closely allied species. Now this astragalus, as we have seen, resembles most closely that of the Megatherium; and since we may infer that the calcaneum, scaphoides, and cuboides had a like correspondence, the inclination of the sole of the foot inwards must have been very slight, as I have determined from examination of the structure and co-adaptation of those bones in the incomplete skeleton of the Megatherium in the London College of Surgeons. Such an inclination of the foot may be conceived to have facilitated the bending of the long claws upon the sole, during the ordinary progressive movements of the animal, but it is quite insufficient to justify the conclusion, that it related to an application of the hind feet for the purposes of climbing.

It is not without interest again to call to mind the deviation of the structure of the astragalus of the Scelidothere from the Megatherioid to the Dasypodoid type of structure. For if the Megatherioid type of structure had really been one suitable to the exigencies of climbing quadrupeds, it might have been expected to have exhibited the scansorial modifications more decidedly, as

---

the Megatherium was really a climber, it must have had still more occasion (on account of its greater size), for that peculiar arrangement of the hind-feet which we have described in the Megalonyx."

the species diminished in stature; but as regards the instructive bone of the hind-foot, the modifications of which we have just been considering, this is by no means the case.

DESCRIPTION OF A MUTILATED LOWER JAW OF THE

MEGALONYX JEFFERSONII.

IN the preceding section an astralagus was described, which was regarded as belonging possibly to the same Edentate species as the jaw figured and described, p. 69, Pl. XVIII. and XIX., under the name of *Mylodon Darwinii;* but the same correspondence,—that of relative size,—renders it equally possible that this astragalus may belong to the species of *Megalonyx* to which the lower jaw now under consideration appertains. There could be no doubt, from its structure, that it was the astragalus of a gigantic species of the order *Bruta*, and of the *Megatherioid* family, and more nearly allied to the Megathere than is the Scelidothere, but sufficiently distinct from both.

The lower jaw, figured in Pl. XXIX., is the only fossil brought home by Mr. Darwin that could be confidently referred to the genus *Megalonyx;* but the form of the tooth in place on the right side of the jaw fully justifies this determination. The jaw itself is deeply and firmly imbedded in the matrix, so that only the upper or alveolar border is visible. The coronoid and condyloid processes are broken away, and the texture of the remaining part of the jaw was too friable, and adhered too firmly to the surrounding matrix to admit of more of its form being ascertained than is figured.

There were four molars on each side of this jaw; the large oblique perforation near the fractured symphysis is the anterior extremity of the wide dental canal. The forms of the alveoli are best preserved in the right ramus: the first is the smallest, and seems to have contained a tooth, of which the transverse section must have been simply elliptical: the second tooth is likewise laterally compressed, but the transverse section is ovate, the great end being turned forwards: the third socket presents a corresponding form, but a larger size: the fourth socket is too much mutilated to allow of a correct opinion being formed as to the shape of the tooth which it once contained. The natural size of the tooth *in situ*, and of the adjoining socket, is given in Pl. XXIX., fig. 2. The difference of form which the jaw of the Megalonyx presents, as compared with that of the Mylodon, especially in the greater recedence of the two horizontal rami from each other, will be appreciated by comparing Pl. XVIII. with Pl. XXIX.

DESCRIPTION OF A FRAGMENT OF THE SKULL AND OF THE TEETH OF THE

MEGATHERIUM CUVIERI.

NOTWITHSTANDING the full, accurate, and elaborate accounts of the skeleton of the Megatherium given by Brû,[*] Cuvier,[†] Pander and D'Alton,[‡]; and Mr. Clift,[§] the fragments of this most gigantic of quadrupeds brought home by Mr. Darwin, possess much interest, and have added, what could hardly have been anticipated, important information as to the dental system, whereby an error in the generic character of the Megatherium

---

[*] Descripcion del Esqueleto de un quadrupedo muy corpulento y raro, que se conserva en el Real Gabinete de Historia Natural de Madrid. Folio, Madrid, 1796.
[†] Ossemens Fossiles, tom. v. pt. i. p. 179.
[‡] Das Riesen Faulthier, Bradypus giganteus, von Dr. Chr. Pander und Dr. E. D'Alton." Folio, Bonn, 1821.
[§] Transactions of the Geological Society, 1835, p. 438.

has been corrected.

The fragments here alluded to are portions of the skull of three full-grown Megatheres: the most perfect part of which affords a view of the posterior, and of part of the basal surface, which regions of the cranium have not hitherto been elsewhere figured or described, (Pl. XXX.)

The plane of the occipital foramen forms with that of the base of the skull an angle of 140°, the plane of the posterior surface of the skull forms with the basal plane an angle of 68°. The occipital condyles are therefore terminal, or form the most posterior parts of the cranium. The extent of their convex curvature in the antero-posterior direction, which equals that of a semicircle, indicates that the Megatherium possessed considerable freedom and extent of motion of the head. The condyles are not extended in the lateral direction so far as in the Toxodon; their axis is more oblique than in the Glossotherium, and their internal surface is more parallel with the axis of the skull, the foramen magnum not presenting that infundibuliform expansion which is so characteristic of the Glossotherium. The occipital condyles resemble most in form and position those of the Scelidotherium; but in the angle of the occipital plane the Megatherium is intermediate between the Scelidothere and Glossothere. The ex-occipitals terminate laterally and inferiorly, each in a short, but strong obtuse process. The posterior plane of the skull is traversed by a strong arched intermuscular crest, which forms the upper boundary of a pretty deep fossa, which is divided by a median vertical ridge, extending downwards to within an inch of the upper margin of the foramen magnum. A second strong obtuse transversely arched ridge curves over the first, and forms the upper boundary of the posterior or occipital region of the skull: the interspace between the two transverse ridges is very irregular, and indicates the firm implantation of powerful nuchal muscles or ligaments, (Pl. XXX. fig. 1.)

In the configuration and angle of the occipital plane the Megatherium indicates the same general correspondence with the Edentate type, which has been pointed out in the descriptions of the crania of the Glossothere and Scelidothere: and the resemblance to the Scelidothere is not less striking in the small proportional size of the cranium in this quadruped, which surpasses the rest of its class in so great a degree in the colossal proportions of its hinder parts.

Having detected in the base of the skull of the Scelidothere an articular semicircular pit for the head of the styloglossal bone, similar to, but relatively smaller than, that remarkable one in the skull of the Glossothere, it became a matter of interest to determine whether this structure, which does not exist in any of the existing Edentals, should likewise be present in the gigantic type of the Megatherioid family. The result of a careful removal of the matrix from the basal region of one of the cranial fragments of the Megatherium was the detection of this articular cavity, in each temporal bone in the same relative position as in the Glossothere and Scelidothere. The styloid articular cavity is relatively smaller, and shallower, than in the Glossothere, its proportions being much the same as those of the Scelidothere. The cranial or posterior extremity of the stylo-hyoid bone in the Scelidotherium is bent upwards at an obtuse angle (Pl. XXI.), and terminates in an articular ball which rotates in this cavity. The size of this bone, and its mode of articulation, indicates great power and muscularity of tongue in the Megatherioids, and calls to mind the importance of that organ in the Giraffe, which subsists on the same kind of food as that which I have supposed to have supported the Megatherioids, although the general organization of these animals and the mode in which the foliage was brought within reach of the tongue are as opposite as can well be imagined.

The anterior condyloid foramen presents scarcely one half the absolute size of that of the Glossothere, whence we may infer a correspondingly inferior development of the tongue in the Megathere. The fractured parietes of the cranial cavity of the Megatherium every where exhibit evidences of the great extent of the air-cells or sinuses continued from the nasal cavity: on the basilar aspect of the cranium they extend as far back as the jugular foramina: the whole of the basi-sphenoid being thus excavated, and permeable to air, derived from the sphenoid sinuses, (Pl. XXX. fig. 2.) The vertical diameter of the cranial cavity is four inches, eight lines; its transverse diameter, which is greatest in the posterior third part of the cavity, corresponding with the posterior part of the cerebrum is six inches: from the indications afforded by the remains of the cranial cavity in Mr. Darwin's specimens, I conclude that the brain of the Megatherium was more depressed, and upon the whole, smaller by nearly one-half than that of the Elephant; but with the cerebellum relatively larger, and situated more posteriorly with relation to the cerebral hemispheres: whence it may be concluded that the Megatherium was a creature of less intelligence, and with the command of fewer resources, or a less varied instinct than the Elephant.

It has been usual to characterize the Megatherium, in conformity with the concurrent descriptions of Bru, Cuvier, and D'Alton, by the dental formula of *molares* 4/4 4/4, i. e. by the presence of four grinding teeth on each side of the upper, as of the lower jaw. It was the agreement of the excellent authorities above cited in this statement, which induced Mr. Clift and myself to regard a single detached tooth, which formed part of the valuable collection of remains of the Megatherium deposited in the Hunterian Museum by Sir Woodbine Parish, as being, from its comparatively small size, the tooth of either a younger individual or of a smaller species of Megatherium. Upon clearing away the matrix from the palatal and alveolar surface of one of the cranial fragments of the Megatherium in Mr. Darwin's collection, I was gratified by the detection of the crown of a fifth molar, corresponding in size and form with the detached tooth, above alluded to: its small size, and its position have doubtless occasioned its being over-looked in the cranium of the great skeleton at Madrid.

The anterior molar of the upper jaw presents a nearly

semicircular transverse section, with the angles rounded off; the three succeeding teeth are four-sided, with the transverse somewhat exceeding the antero-posterior diameter: they are rather longer and larger than the first: the last molar is likewise four-sided, but presents a sudden diminution of diameter, and is relatively broader. The following are the respective dimensions of the upper maxillary teeth.

	First Molar.		Second Molar.		Third Molar.		Fourth Molar.		Fifth Molar.	
	In.	Lines.	In.	Lines.	In.	Lines.	In.	Lines.	In.	Lines.
Length	8	6	9	4	9	4	8	7	5	2
Transverse diameter..	1	9	2	4	2	3	2	0	1	4
Antero-posterior diameter..	1	5	2	0	2	0	1	11	0	10

Besides the differences in size, the upper molars vary as to their curvature: this difference is exhibited in the vertical section of these teeth figured in Pl. XXXI. The convexity of the curve of the first, second and third molars is directed forwards; the fourth is straight, its anterior surface only describing a slight convexity in the vertical direction; the fifth tooth is curved, but in a contrary direction to the others; and the bases of the five molars thus present a general convergence towards a point a little way behind the middle of the series.

The next peculiarity to be noticed in these remarkable teeth is the great length of the pulp-cavity (d), the apex of which is parallel with the alveolar margin of the jaw: a transverse fissure is continued from this apex to the middle concavity of the working surface of the tooth, which is thus divided into two parts. Each of these parts consists of three distinct substances,—a central part analogous to the body or bone of the tooth or 'dentine,' a peripheral and nearly equally thick layer of *cæmentum*, and an intermediate thinner stratum of a denser substance, which is described in Mr. Clift's memoir on the Megatherium as 'enamel,' and to which substance in the compound teeth of the Elephant, it is analogous both in its relative situation, and relative density to the other constituents.

Microscopic examinations of thin and transparent slices of the tooth of the Megatherium prove, however, that the dense layer separating the internal substance from the cæmentum is not enamel, but presents the same structure as the hard 'dentine' or ivory of the generality of Mammalian teeth; and corresponds with the thin cylinder of hard 'dentine' in the tooth of the Sloth. No species of the Order *Bruta* has true enamel entering into the composition of its teeth; but the modifications of structure which the teeth present in the different genera of this order are considerable, and their complexity is not less than that of the enamelled teeth of the Herbivorous Pachyderms and Ruminantia, in consequence of the introduction of a dental substance into their composition corresponding in structure with that of the teeth of the *Myliobates*, *Psammodus*, and other cartilaginous fishes.

The microscopic investigation of the structure of the teeth of the Megatherium was undertaken chiefly with the view of comparing this structure with that of the teeth of the Sloth and Armadillo, and of thus obtaining an insight into the food, and an additional test of the real nature of the disputed affinities of the Megatherium. The central part of the tooth (*c.* Pl. XXXI.) consists of a coarse ivory, like the corresponding part of the tooth of the Sloth. It is traversed throughout by medullary canals 1/1500th of an inch in diameter, which are continued from the pulp-cavity, and proceed, at an angle of 50°, to the plane of the dense ivory, parallel to each other, with a slightly undulating course, having regular interspaces, equal to one and a half diameters of their own arcæ, and generally anastomosing in pairs by a loop of which the convexity is turned towards the origin of the tubes of the fine dentine, as if each pair so joined consisted of a continuous reflected canal, (*c.* fig. 1, Pl. XXXII.) The loops are generally formed close to the fine dentine. In a few situations I have observed one of the medullary canals continued across the fine dentine, and anastomosing with the corresponding canals of the cæmentum. The interspaces of the medullary canals of the coarse dentine are principally occupied by calcigerous tubes which have an irregular course, anastomose reticularly, and terminate in very fine cells. The more regular and parallel calcigerous tubes, which constitute the thin layer of hard dentine, are given off from the convexity of the terminal loops of the medullary canals. The course of these tubes (b. fig. 1, Pl. XXXII.) is rather more transversely to the axis of the tooth than the medullary canals from which they are continued. They run parallel to each other, but with minute undulations throughout their course, in which they are separated by interspaces equal to one and a half their own diameter. As they approach the cæmentum they divide and sub-divide, and grow more wavy and irregular: their terminal branches take on a bent direction, and form anastomoses, dilate into small cells, and many are seen to become continuous with the radiating fibres or tubes of the cells or corpuscles of the contiguous cæmentum. This substance enters largely into the constitution of the compound tooth of the Megatherium: it is characterized, like the cæmentum of the Elephant's grinder, by the presence of numerous radiated cells, or purkingian corpuscles, scattered throughout its substance, but may be distinguished by wide medullary canals which traverse it in a direction parallel with each other, and forming a slight angle with the transverse axis of the tooth. These canals are wider than those of the central coarse dentine, their diameter being 1/1200th of an inch; they are separated by interspaces equal to from four to six of their own diameters, divide a few times dichotomously in their course, and finally anastomose in loops, the convexity of which is directed towards, and in most cases is in close contiguity with, the layer of dense dentine.

Fine calcigerous tubes are every where given off at right angles from the medullary canals of the cæmentum, which form a rich reticulation in their interspaces, and a direct continuation between the loops of the medullary canals and the calcigerous tubes of the dense dentine. The cæmentum differs from the coarse dentine in the larger size and wider interspaces of its medullary canals, and by the presence of the bone-corpuscles in their interspaces; but they are brought into organic communication with each other, not only by means of the tubes of the dense dentine, but by occasional continuity of the medullary canals across that substance. The tooth of the Megatherium thus offers an unequivocal example of a course of nutriment from the dentine to the cæmentum, and reciprocally. Retzius observes with respect to the human tooth, that "the fine tubes of the cæmentum enter into immediate communications with the cells and tubes of the dentine (zahnknochen), so that this part can obtain from without the requisite humours after the central pulp has almost ceased to exist." In the Megatherium, however, those anastomoses have not to perform a vicarious office, since the pulp maintains its full size and functional activity during the whole period of the animal's existence. It relates to the higher organized condition, and greater degree of vitality of the entire grinder in that extinct species.

The conical cavities (*d.* Pl. XXXI.) attest the size and form of the persistent pulp; the diameter of its base is equal to the part of the crown of the tooth which is formed by the coarse and fine dentine. From the gradual thinning off, and final disappearance of these substances as they reach the base of the tooth, I conclude that they were both formed at the expense of the pulp. The fine tubes and cells must have been excavated in its peripheral layer for the reception of the hardening salts of the dense dentine, and the rest converted into the parallel series of medullary canals with their respective systems of calcigerous tubes, in a manner closely analogous to the development of the entire tooth of the Orycteropus. The coarser dentine of the tooth of the Megatherium differs, in fact, from the entire tooth of the *Orycteropus*, only in that the parallel medullary canals and their radiating calcigerous tubes are not separated from the contiguous canals by a distinct layer of cæmentum, and that the medullary canals anastomose at their peripheral extremities. The wide spaces, (*e.* Pl. XXXI.)

indicate the thickness of the dental capsule by the ossification of which the exterior stratum of cement was formed. It was not until I knew the true structure of the tooth of the Megatherium, that I could comprehend the mode of its formation. The parallel layers of enamel in the Elephant's grinder are formed, as is well known, by membranous plates passing from the coronal end of the closed capsule towards the base of the tooth; but a certain extent of enamel can only thus be formed, and when the crown of the grinder has once protruded, and come into use, the enamel cannot be added to. The modification of the structure of the tooth of the Megatherium readily permits the uninterrupted and continuous formation of the dense substance which is analogous to the enamel of the Elephant's grinder.

With respect to the question of the respective affinities of the Megatherium to the Bradypodoid or Dasypodoid families, the result of this examination of the teeth speaks strongly for its closer relationship with the former group: the Megalonyx, Mylodon, and Scelidotherium, in like manner correspond in the structure of their teeth with the Sloth, and differ from the Armadillo.

If from a similarity of dental structure we may predicate a similarity of food, it may reasonably be conjectured that the leaves and soft succulent sprouts of trees may have been the staple diet of the Megatherioid quadrupeds, as of the existing Sloths. Their enormous claws, I conclude, from the fossorial character of the powerful mechanism by which they were worked, to have been employed, not, as in the Sloths, to carry the animal to the food, but to bring the food within the reach of the animal, by uprooting the trees on which it grew.

In the remains of the Megatherium we have evidence of the frame-work of a quadruped equal to the task of undermining and hawling down the largest members of a tropical forest. In the latter operation it is obvious that the immediate application of the anterior extremities to the trunk of the tree would demand a corresponding fulcrum, to be effectual, and it is the necessity for an adequate basis of support and resistance to such an application of the fore-extremities which gives the explanation to the anomalous development of the pelvis, tail, and hinder extremities in the Megatherioid quadrupeds. No wonder, therefore, that their type of structure is so peculiar; for where shall we now find quadrupeds equal, like them, to the habitual task of uprooting trees for food?

DESCRIPTION OF FRAGMENTS OF BONES, AND OF OSSEOUS TESSELATED DERMAL COVERING OF LARGE EDENTATA.

It is now determined that there once existed in South America, besides the Megatherium, the Megalonyx, and the allied genera described in the preceding pages of the present work, gigantic species of the order *Bruta* belonging to the Armadillo family, and defended, like the small existing representatives of that family, by a tesselated bony dermal covering. The largest known species of these extinct *Dasypodidœ* is the *Glyptodon clavipes*, of which the armour and parts of the skeleton have been described by MM. Weiss and D'Alton in the Berlin Transactions for 1827 and 1834: and the generic and specific characters and name, with an account of the dental system, and bones of the extremities, were recorded in the Geological Proceedings for March 1839. It would seem that parts of the same, or a nearly allied gigantic species were described in the same year by M. Lund; under the name of *Hoplophorus*. Of the valuable and interesting discoveries of this able Naturalist I regret that I was not aware until the appearance of a notice of them in the Comptes Rendus for April, 1839.[*] Amongst the fragments of bony tesselated armour in Mr.

---

[*] An excellent translation of the description of the Brazilian fossils found by M. Lund, is published in the Annals of Natural History, July and August, 1839.

Darwin's collection are a few pieces which were found by him, associated with remains of Toxodon and Glossotherium near the Rio Negro in Banda Oriental.[†] These fragments, if we may judge from their thickness, must have belonged to an animal at least as large as the *Glyptodon clavipes;* but the pattern differs in the greater equality of size of the component tesseræ. The thickness of the largest fragment is one inch and a half, the tesseræ vary in diameter from one inch to half an inch, and are separated by grooves about two lines in depth, and two in diameter. The pattern formed by the anastomosis of these grooves is an irregular net-work; the contour of the tesseræ is either unevenly subcircular, hexagonal, pentagonal, or even four-sided; with the sides more or less unequal. In those portions of this armour, where one of the tesseræ exceeds the contiguous ones in size, the imagination may readily conceive it to be the centre of a rosette, around which the smaller ones arrange themselves, but there is no regular system of rosettes, as in the portions of the dermal armour of the Glyptodon figured by Weiss, and those brought to England by Sir Woodbine Parish, in which the central piece is double the size of the marginal ones.

The portions of the tesselated bony dermal covering of a Dasypodoid quadruped, figured in Pl. XXXII. figs. 5 and 4, of the natural size, were discovered folded round the middle and ungueal phalanges, figs. 2 and 3, at Punta Alta, in Bahia Blanca, in an earthy bed interstratified with the conglomerate containing the remains of the fossil Edentals.

In one of these fragments, measuring six inches long by five broad, the tesseræ are arranged in rosettes, and so closely correspond in size and pattern with the bony armour described by M. Lund, as characterizing his species, *Hoplophorus euphractus*, that I feel no hesitation in referring them to that animal. One of

---

[†] At the distance of a few leagues from the locality here mentioned, other fragments were found by Mr. Darwin; also near Santa Fé, in Entre Rios; also on the shores of the Laguna, near the Guardia del Monte, South of Buenos Ayres; also, according to the Jesuit Falkner, on the banks of the Tercero.

the pattern rosettes is figured at fig. 4, together with the thickness of the armour at this part, and the coarse tubulo-cellular structure of the bone. Another portion of dermal armour from the same locality, gives the pattern shown in fig. 5, formed by square or pentagonal tesseræ, arranged in transverse rows; it is certain that this portion of armour belonged to the same animal as the preceding piece; and probably that it constituted part of the transverse dorsal bands of the *Hoplophorus*.

The middle and ungueal phalanx, as well as the portions of armour, are given of the natural size, in Pl. XXXII. The upper and outer surface of the phalanx, is shown in fig. 2. It is smooth and flat; joins the inner surface by a sharp edge, which runs along the upper and inner side of the bone; and passes by a gradual convexity to the under surface; the ridge corresponding with the base of the claw, is feebly developed at the under and lateral parts of the base of the claw. Below the double trochlear joint for the middle phalanx, there are two articular surfaces for two large sesamoid bones.

The middle phalanx corresponds in its small antero-posterior diameter and wedge-shape, with that of the great Glyptodon: but the terminal phalanx is longer and deeper, in proportion to its breadth.

Among the collection of fossils from Punta Alta, in Bahia Blanca, there is an interesting fragment of the head of a gigantic animal of the Edentate order, including the glenoid cavity, and part of the zygomatic process of the left side. The articular surface for the lower jaw, exhibits, in its flatness, extent, and the absence of a posterior ridge, the well-marked characteristics of this part of the Edental structure. It measures two inches four lines in the transverse, and two inches two lines in the antero-posterior diameter. The commencement of the zygomatic process presents a vertical diameter of two inches, and a transverse diameter of eight lines at the thickest part. It is slightly concave at its lower border, and convex above. The small portion of the cranial parietes, which is preserved, exhibits the cellular structure consequent upon the great extension and development of the nasal air-sinuses: this condition of the cranial parietes, has already been noticed in the description of the more perfect skulls of the large extinct Edentata.

NOTICE OF FRAGMENTS OF MOLAR TEETH OF A

MASTODON.

Of the remains of this gigantic extinct Pachyderm, observed by Mr. Darwin at Santa Fé, in Entre Rios, and on the banks of the Tercero, the fragments of the teeth and portions of the skeleton which reached England, are not sufficient to lead to a determination of the species; but sufficiently prove it to have been nearly allied, if not identical, with the *Mastodon angustidens* of Cuvier, and unquestionably distinct from the *Mastodon giganteum* of the United States.

NOTICE OF THE REMAINS OF A SPECIES OF

EQUUS,

Found associated with the extinct Edentals and Toxodon at Punta Alta, in Bahia Blanca, and with the Mastodon and Toxodon at Santa Fé, in Entre Rios.

The first of these remains is a superior molar tooth of the right side; it was embedded in the quartz shingle, formed of pebbles strongly cemented together with calcareous matter, which adhered as closely to the tooth in question, as the corresponding matrix did to the associated fossil remains. The tooth was as completely fossilized as the remains of the Mylodon, Megatherium, and Scelidothere; and was so far decomposed, that in the attempt to detach the adherent matrix, it became partially resolved into its component curved lamellæ. Every point of comparison that could be established proved it to differ from the tooth of the common *Equus Caballus* only in a slight inferiority of size.

The second evidence of the co-existence of the horse with the extinct Mammals of the tertiary epoch of South America reposes on a more perfect tooth, likewise of the upper jaw, from the red argillaceous earth of the Pampas at Bajada de Santa Fé, in the Province of Entre Rios.*

This tooth is figured at Pl. XXXII. fig. 13 and 14, from which the anatomist can judge of its close correspondence with a middle molar of the left side of the upper jaw.

This tooth agreed so closely in colour and condition with the remains of the Mastodon and Toxodon, from the same locality, that I have no doubt respecting the contemporaneous existence of the individual horse, of which it once formed part.

This evidence of the former existence of a genus, which, as regards South America, had become extinct, and has a second time been introduced into that Continent, is not one of the least interesting fruits of Mr. Darwin's palæontological discoveries.

---

* Mr. Darwin has more particularly described the circumstances of the embedment of this tooth in his Journal of Researches, p. 149, during the Voyage of the Beagle.

DESCRIPTION OF REMAINS OF RODENTIA, INCLUDING THE JAWS AND TEETH OF AN EXTINCT SPECIES OF

CTENOMYS.

The fragment of the upper jaw, figured in Pl. XXXII. fig. 6, exhibits the first and second molar *in situ*, and the socket of the third and fourth molar, of a Rodent, which by the form and number of the upper maxillary teeth is referable to the genus Ctenomys. The molars are a little larger, the longitudinal groove on their external surface is somewhat deeper, and the last molar is relatively wider than in the existing subterraneous species,—the Tucutucu (*Ctenomys Brasiliensis*, Bl.), of whose habits so interesting an account is given in the description of the Mammalia of the present Collection (No. IV. p. 79). The form of the grinding surface of the first and second upper molar is shown below the fig. 6, and three views of the second grinder are given at figs. 7, 8, and 9. The fragment of the lower jaw of the same fossil Rodent is figured at fig. 10 and 11. The long anterior incisor is relatively narrower than in the *Ctenomys Brasiliensis*. I have not had the means of comparing this fossil with the *Ctenomys Magellanicus;* but since it is probable that the *Ct. Magellanicus* may not be specifically different from the *Ct. Brasiliensis*, it may be concluded that the present fossil is equally distinct from both.

The portion of the right hind-foot of the Rodent figured at fig. 12, includes the calcaneum, astragalus, cuboides, external and middle cuneiform bones, and the metatarsals and proximal phalanges of the toes corresponding with the three middle toes of five-toed quadrupeds. The metatarsals are chiefly remarkable for the well-developed double-trochlear articular surface, and intermediate ridge. These remains, as well as the jaws and teeth of the Ctenomys, were discovered at Monte Hermoso in

Bahia Blanca.

In the same reddish earthy stratum of that locality, Mr. Darwin discovered the decomposed molar of a Rodent, equalling in size, and closely resembling in the disposition of its oblique component laminæ, the hinder molar of the Capybara (*Hydrochærus.*) The fossil differs, however, in the greater relative breadth of the component laminæ.

I have, lastly, to notice the head of a femur, and some fragments of pelvic bones from the same formation which bear the same proportion to the tooth above alluded to, as subsists between the teeth and bones of the Capybara, and which are sufficient to prove that there once has existed in South America a species of the family *Caviidæ*, as large as the present *Capybara*, but now apparently extinct.

This fact, together with the greater part of those which have been recorded in the foregoing pages of the present work, establishes the correspondence, in regard to the characteristic type, which exists between the present and extinct animals of the South American Continent: we have abundant evidence likewise of the greater number of generic and specific modifications of these fundamental types which the animals of a former epoch exhibited, and also of the vastly superior size which some of the species attained.

At the same time it has been shewn that some of the present laws of the geographical distribution of animals would not have been applicable to South America, at the period when the Megatherioids, Toxodon, and Macrauchenia existed: since the Horse, and according to M. Lund, the Antelope and the Hyæna, were then associated with those more strictly South American forms. The Horse, which, as regards the American continent, had once become extinct, has again been introduced, and now ranges in countless troops over the pampas and savannahs of the new world. If the small Opossums of South America had been in like manner imported into Europe, and were now established like the Squirrels and Dormice in the forests of France, an analogous case would exist to that of the Horse in South America, as the fossil Didelphys of Montmartre proves with respect to the geological contemporaneity of the fossils collected by him, Mr. Darwin subjoins the following observations:—

"The remains of the following animals were embedded together at Punta Alta in Bahia Blanca:—The *Megatherium Cuvierii, Megalonyx Jeffersonii, Mylodon Darwinii, Scelidotherium leptocephalum, Toxodon Platensis* (?) a Horse and a small Dasypodoid quadruped, mentioned p. 107; at St. Fé in Entre Rios, a Horse, a Mastodon, *Toxodon Platensis*, and some large animal with a tesselated osseous dermal covering; on the banks of the Tercero the Mastodon, Toxodon, and, according to the Jesuit Falkner, some animal with the same kind of covering; near the Rio Negro in Banda Oriental, the *Toxodon Platensis*, Glossotherium, and some animal with the same kind of covering. To these two latter animals the *Glyptodon clavipes*, described by Mr. Owen in the Geological Transactions, may, from the locality where it was discovered, and from the similarity of the deposit which covers the greater part of Banda Oriental, almost certainly be added, as having been contemporaneous. From nearly the same reasons, it is probable that the Rodents found at Monte Hermoso in Bahia Blanca, co-existed with the several gigantic mammifers from Punta Alta. I have, also, shown in the Introduction, that the *Macrauchenia Patachonica,* must have been coeval, or nearly so, with the last mentioned animals. Although we have no evidence of the geological age of the deposits in some of the localities just specified, yet from the presence of the same fossil mammifers in others, of the age of which we have fair means of judging, (in relation to the usual standard of comparison, of the amount of change in the specific forms of the invertebrate inhabitants of the sea,) we may safely infer that *most* of the animals described in this volume, and likewise the Glyptodon, were strictly contemporaneous, and that *all* lived at about the same very recent period in the earth's history. Moreover, as some of the fossil animals, discovered in such extraordinary numbers by M. Lund in the caves of Brazil, are identical or closely related with some of those, which lately lived together in La Plata and Patagonia, a certain degree of light is thus thrown on the antiquity of the ancient Fauna of Brazil, which otherwise would have been left involved in complete darkness."

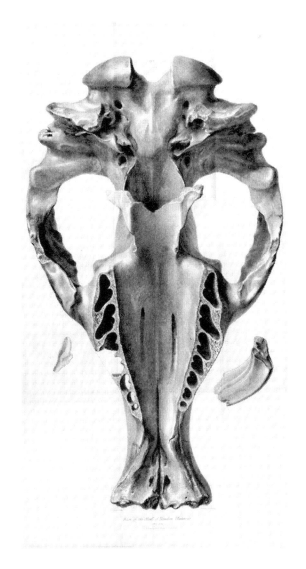

G. Scharf del et lithog.   Printed by C. Hallmandel.
Base of the Skull of Toxodon Platensis.
Nat: Size.

THE ZOOLOGY OF THE VOYAGE OF H.M.S. BEAGLE

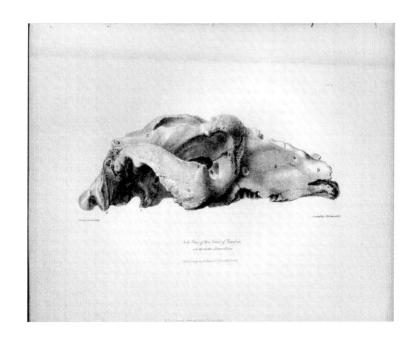

G. Scharf del et lithog.   Printed by C Hallmandel.
Side View of the Skull of Toxodon
one third the Natural Size.

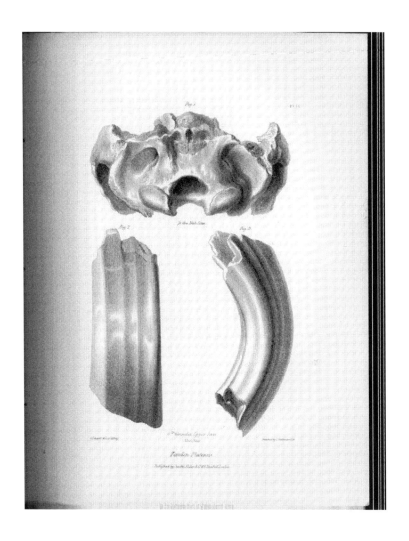

G. Scharf del et lithog.   Printed by C Hallmandel.
Top View of the Skull of the Toxodon.
One third the Nat: Size.

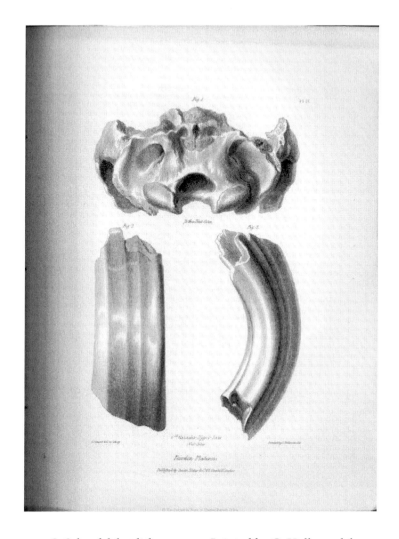

G. Scharf del et lithog.   Printed by C. Hallmandel.
6th Grinder—Upper Jaw
Nat: Size.
Toxodon Platensis.

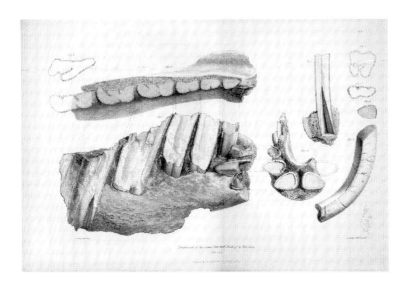

G. Scharf del et lithog.   Printed by C. Hallmandel.
Fragments of the lower Jaw and Teeth of a Toxodon.
Nat Size.

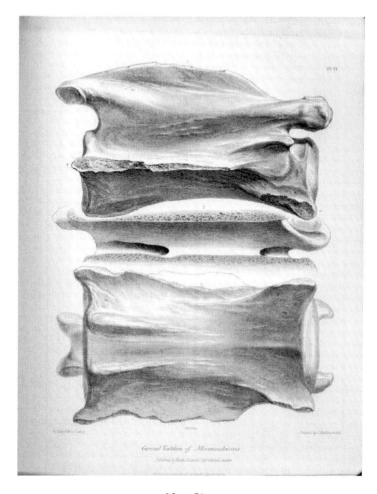

Nat: Size.
G. Scharf del et lithog.   Printed by C. Hallmandel.
Cervical Vertebræ of Macrauchenia.

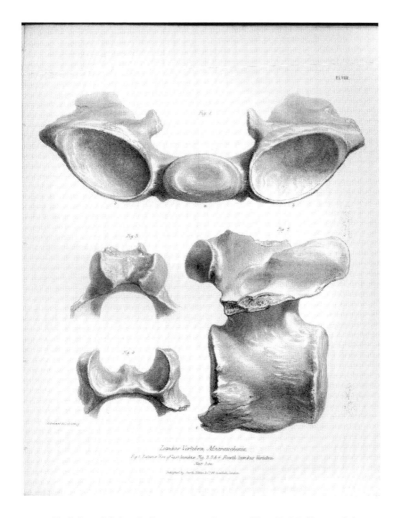

G. Scharf del et lithog.   Printed by C. Hallmandel
Nat: Size.
Cervical Vertebræ of
1.2. Macrauchenia. 3.4. Auchenia.

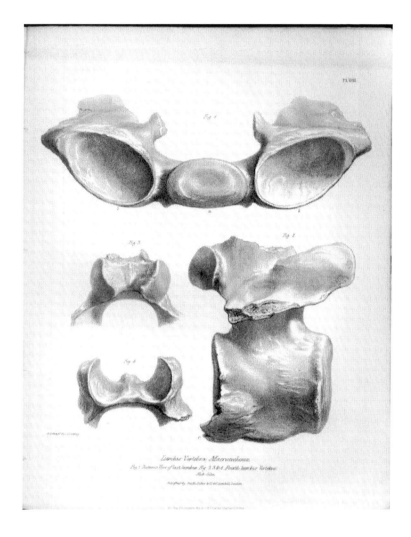

G. Scharf del et lithog.
Lumbar Vertebræ; Macrauchenia.
Fig. 1. Posterior View of last lumbar.
Fig: 2. 3 & 4. Fourthlubar Vertebra.
Nat: Size.

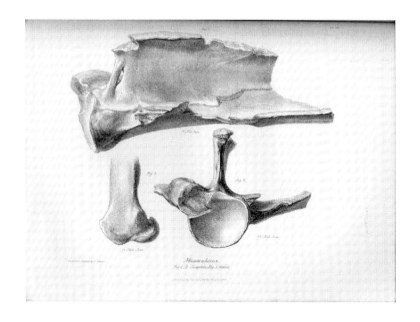

Lithog from Nature by G. Scharf.
Macrauchenia.
Fig. 1-2. Scapula Fig. 3. Femur.

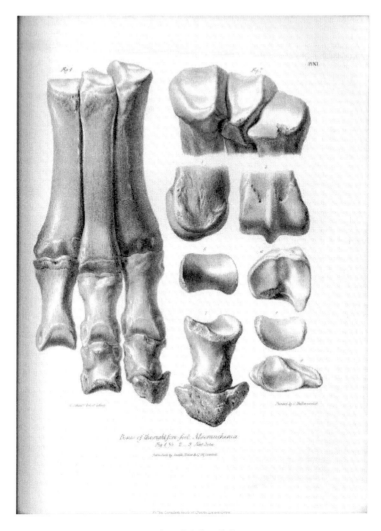

G. Scharf del et lithog.
Proximal Extremity of anchylosed Ulna and Radius
Macrauchenia.
2/3 Nat. Size.

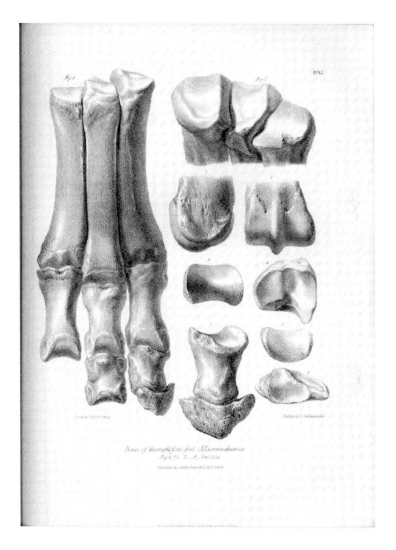

G. Scharf del et lithog.   Printed by C. Hallmandel
Bones of the right fore-foot, Macrauchenia.
Fig 1, 2/3. 2-9, Nat. Size.

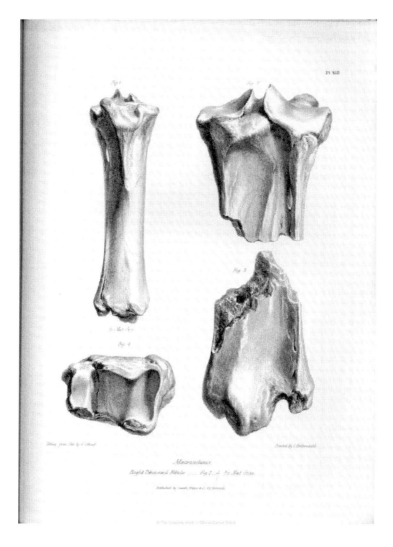

Lithog from Nat by G. Scharf.   Printed by C. Hallmandel.
Right Femur. Macrauchenia.

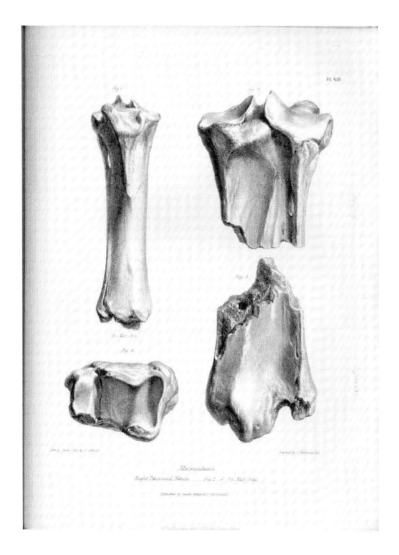

Lithog from Nat. by G. Scharf.   *Printed by C Hallmandel.*
Macrauchenia.
Right Tibia and Fibula. — Fig. 2-4 2/3 Nat. Size.

THE ZOOLOGY OF THE VOYAGE OF H.M.S. BEAGLE

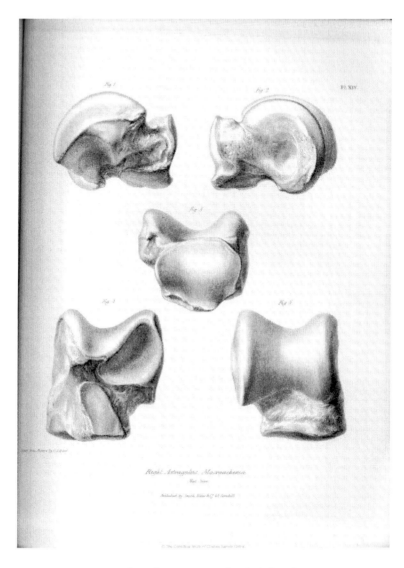

Lithog from Nature by G. Scharf.
Right Astragalus. Macrauchenia.
Nat. Size.

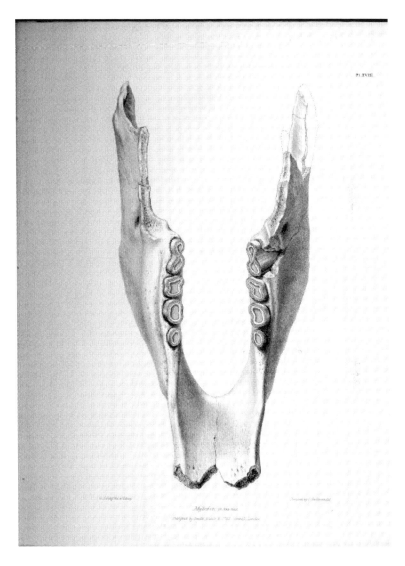

Lithog from Nature by G. Scharf.   *Printed by C. Hallmandel.*
Macrauchenia.
Fig: 1 Metatarsal. 2-5. Metacarpals. Nat. Size.

THE ZOOLOGY OF THE VOYAGE OF H.M.S. BEAGLE

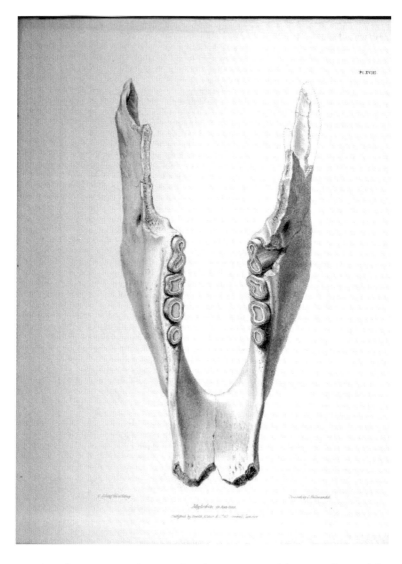

Lithog from Nature by G. Scharf.   Printed by C Hallmandel.
Fragment of the Cranium of the Glossotherium.
1/2 Nat. Size.

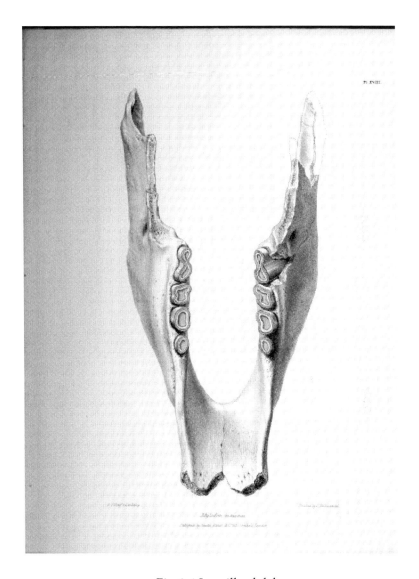

Fig. 3,4 Laurillard del.
Fig 5 G. Scharf del et lithog.
Printed by C Hallmandel.
1. Megalonyx Jeffersoni. 2. Meg laqueatus.
3,4 Mylodon Harlani. 5. Myl Darwinii.

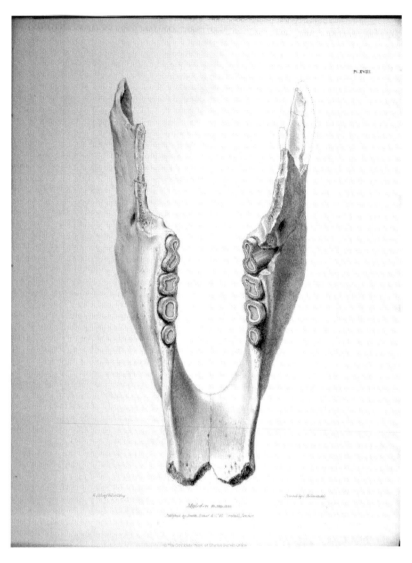

G. Scharf del et lithog. Printed by C. Hallmandel.
Mylodon: 5/9 Nat. Size.

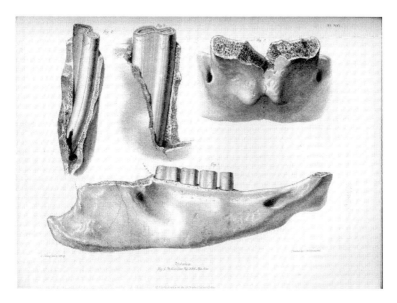

G. Scharf del et lithog.   Printed by C. Hallmandel.
Mylodon.
Fig: 1. 5/9 Nat: Size. Fig: 2.3.4. Nat: Size.

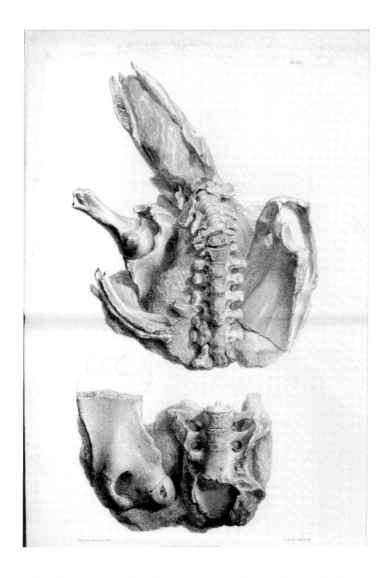

*Lithog. from Nature by G. Scharf.   Printed by C. Hallmandel.Scelidotherium.
1/3 Nat. Size.*

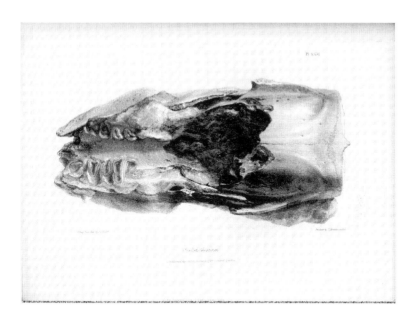

G. Scharf del et lithog.   Printed by C. Hallmandel.
Scelidotherium.
Fig: 1. & 2. 2/3 Nat: Size. Fig: 3-5. Nat: Size.

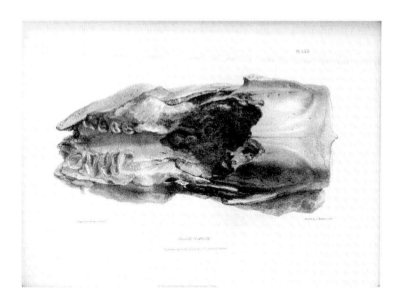

Lithog. from Nat: by G. Scharf.   Printed by C. Hallmandel.
Scelidotherium.

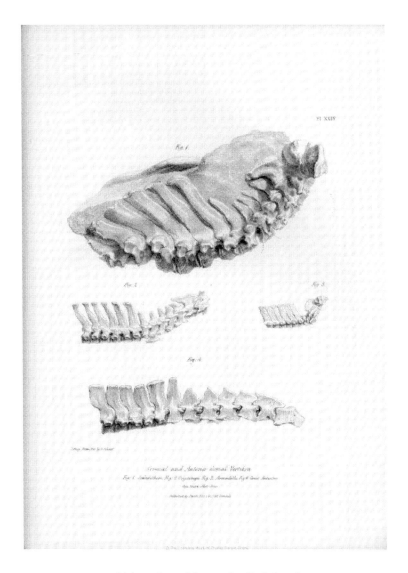

Lithog. from Nature by G. Scharf.
Cranial Cavity and Dentition of Scelidotherium.
Nat: Size.

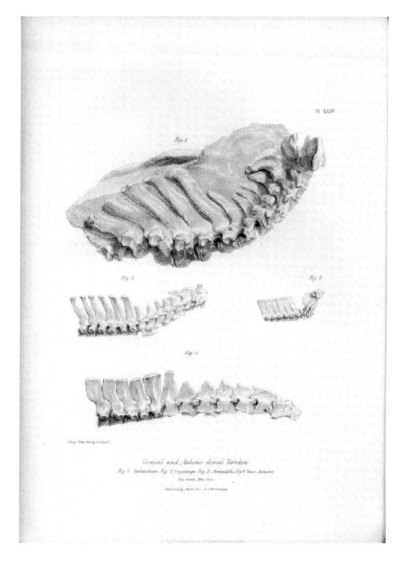

Lithog: from Nat: by G. Scharf.
Cervical and Anterior dorsal Vertebræ.
Fig: 1. Scelidothere. Fig: 2. Orycterope.
Fig: 3. Armadillo. Fig: 4. Great Anteater.
One third Nat: Size.

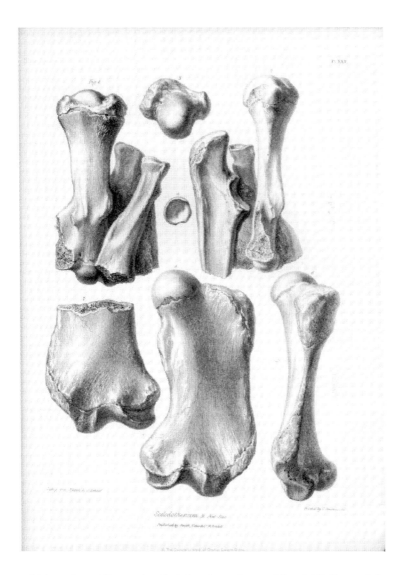

Lithog. from Nature by G. Scharf.   Printed by C. Hallmandel.
Scelidotherium. 1/3 Nat: Size.

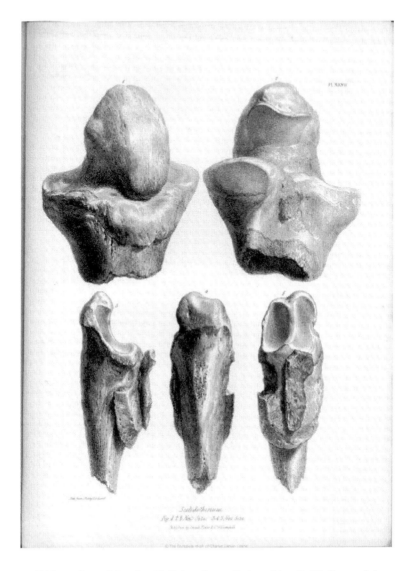

Lithog: from Nat: by G. Scharf.   Printed by C. Hallmandel.
Left Astragalus.
Fig 1.3.5. Megatherium: 1/3 Nat: Size. 2.4.6. Scelidotherium.
2/3 Nat: Size.

FOSSIL MAMMALIA

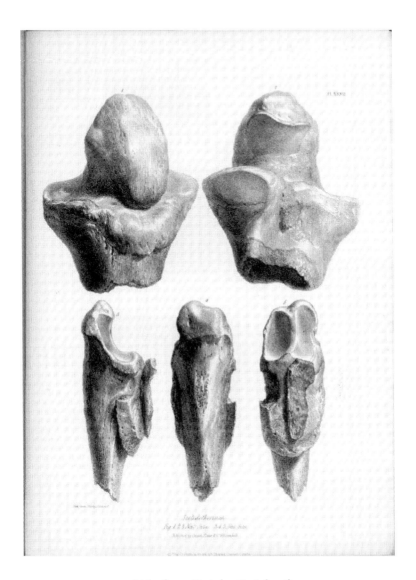

Lith. from Nat. by G. Scharf.
Scelidotherium.
Fig: 1. 2. 2/3 Nat: Size. 3.4.5 Nat: Size.

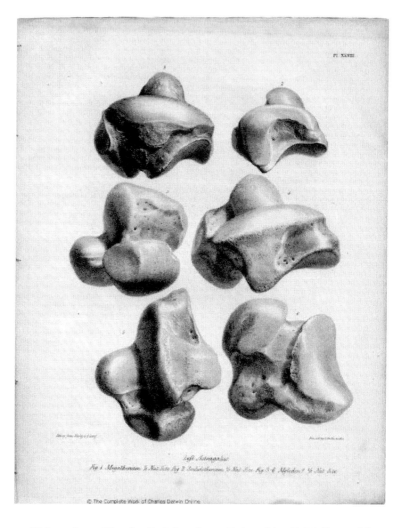

Lithog. from Nat. by G. Scharf.   *Printed by C. Hallmandel.*
Left Astragalus.
Fig. 1. Megatherium. 1/3 Nat. Size. Fig. 2. Scelidotherium. 2/3 Nat. Size. Fig. 3-6. Mylodon. ? 2/3 Nat. Size.

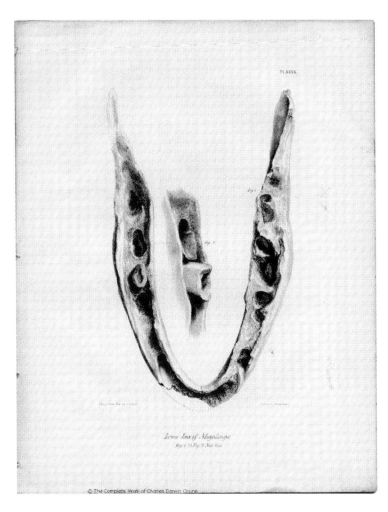

Lithog. from Nat. by G. Scharf.   Printed by C. Hallmandel.
Lower Jaw of Megalonyx.
Fig: 1. 2/3 Fig: 2. Nat: Size.

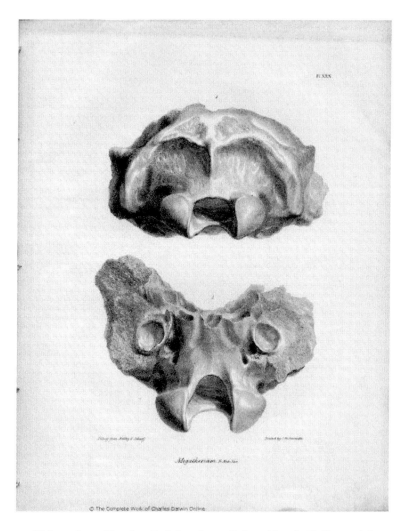

Lithog. from Nat. by G. Scharf.   Printed by C. Hallmandel.
Megatherium. 1/2. Nat: Size.

FOSSIL MAMMALIA

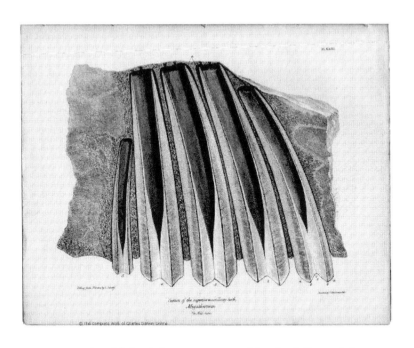

G. Scharf del et lithog.   Printed by C. Hallmandel.
Section of the superior maxillary teeth,
Megatherium.
3/4 Nat. Size.

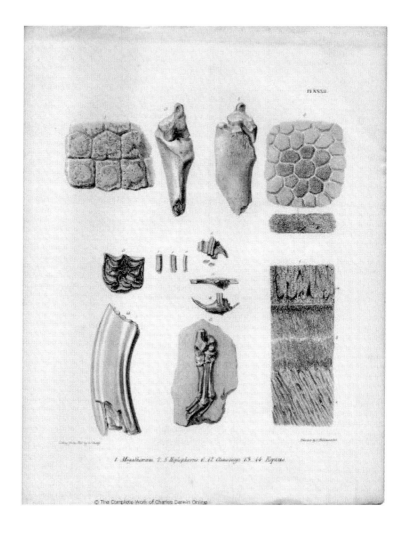

Lithog. from Nat: by G. Scharf.   Printed by C. Hallmandel.
1. Megatherium. 2-5 Hoplophorus. 6-12. Ctenomys.
13-14. Equus.

FOSSIL MAMMALIA

Printed in Great Britain
by Amazon

72722436R00123